Innovating STEM Education

Increased Engagement and Best Practices

Edited by

Eugenia Koleza, Christos Panagiotakopoulos, Constantine Skordoulis

Innovating STEM Education

Increased Engagement and Best Practices

Edited by

Eugenia Koleza, Christos Panagiotakopoulos,
Constantine Skordoulis

First published in 2022
as part of the *The Learner* Book Imprint
doi: 10.18848/978-1-86335-251-2/CGP (Full Book)

Common Ground Research Networks
2001 South First Street, Suite 202
University of Illinois Research Park
Champaign, IL
61820

Copyright © Eugenia Koleza, Christos Panagiotakopoulos, Constantine Skordoulis, 2022

All rights reserved. Apart from fair dealing for the purposes of study, research, criticism or review as permitted under the applicable copyright legislation, no part of this book may be reproduced by any process without written permission from the publisher.

Library of Congress Cataloging-in-Publication Data

Names: Koleza, Eugenia, editor. | Panagiotakopoulos, Christos, editor. | Skordoulis, Constantine, editor.
Title: Innovating STEM education: increased engagement and best practices / [edited by] Eugenia Koleza, Christos Panagiotakopoulos, Constantine Skordoulis.
Description: Champaign, IL: Common Ground Research Networks, 2021. | "This volume contains selected papers presented at the Hellenic Conferences 'Innovating STEM education HiSTEM 2016 and 2018' organized by the Postgraduate Program 'Interdisciplinary Approach on Science, Technology, Engineering and Mathematics in Education STEM Education'"--Introduction. | Includes bibliographical references. | Summary: "In recent years, there has been a focus on promoting the uptake of STEM subjects in schools. This has been driven by the need to ensure that young people gain the knowledge and skills essential to help them participate in a society in which mathematics, science and technology are increasingly important. Nevertheless, reform efforts, including curriculum development, have treated the STEM subjects mostly in isolation. Recognizing that efforts for education within each individual STEM discipline would encourage a wide range of conservations about different important aspects of teaching and learning, this conference considered the potential benefits and challenges for the integration of various STEM's characteristics into education. In order to prepare students to address the problems of our society, it is necessary to provide them with opportunities to understand these problems through rich, engaging and powerful experiences that integrate the disciplines of STEM"-- Provided by publisher.
Identifiers: LCCN 2021028433 (print) | LCCN 2021028434 (ebook) | ISBN 9781863352499 (hardback) | ISBN 9781863352505 (paperback) | ISBN 9781863352512 (adobe pdf)
Subjects: LCSH: Science--Study and teaching--Congresses. | Mathematics--Study and teaching--Congresses. | Interdisciplinary approach in education--Congresses.
Classification: LCC Q181 .I6527 2021 (print) | LCC Q181 (ebook) | DDC 507.1--dc23
LC record available at https://lccn.loc.gov/2021028433
LC ebook record available at https://lccn.loc.gov/2021028434

Table of Contents

Introduction 1

    Eugenia Koleza, Christos Panagiotakopoulos, & Constantine Skordoulis

Chapter 1 9
Describing the Transition from STEM to STEAM by Replicating a Historical Scientific Instrument
    Vasiliki Psoma[1] & Constantine Skordoulis[1]
    1. Introduction
    2. Theoretical Framework
    3. Educational Approach: EDP and Project-Based Learning (PBL)
    4. Discussion

Chapter 2 19
The Integration of STEM Curriculum in the Design of A STEM Scenario
    John Sdrallis[1] & Eugenia Koleza[1]
    1. Introduction
    2. Teaching Natural Science in STEM-Based Scenarios
    3. "Integrating" The Curricula
    4. Biodiesel": A Teaching Proposal

Chapter 3 29
STEMART = STEM + ART: An Erasmus+ STEAM Project for K-12 Education
    Milan Hausner[1], Paraskevi Karypidou[2], Maria Gratiela Popescu[3], Serkan Altinöz[4], Laura Bozzo[5], Olya Mihova[6], & Petra Mikše[7]
    1. Introduction
    2. Goals and Methodology

3. Implementation and Outcomes
4. Impact
5. Dissemination and Use of Results

Chapter 4 41
Searching for Black Holes. Photometry in Our Classrooms.
    Ioannis Chiotelis[1] & Maria Theodoropoulou[2]
    1. Introduction
    2. Experiment
    3. Results

Chapter 5 53
Developing a Design Framework for UMI Educational Scenarios
    Olga Fragou[1], Achilleas D. Kameas[1,2], & Ioannis D. Zaharakis[1,3]
    1. Introduction
    2. U-learning
    3. Developing a Design Framework for U-Learning Educational
    4. Methodology
    5. Towards Developing UMI-Sci-Ed Educational Scenarios Smes Case Study I
    6. Conclusions

Chapter 6 65
New Perspectives for Geometry Teaching: Mechanical Linkages Technology
    Kalliopi Siopi[1] & Eugenia Koleza[2]
    1. Introduction
    2. Machines and Mechanisms
    3. Working with a Pantograph in A Secondary School
    4. Some Remarks

Chapter 7 77
Introducing STEM to Primary Education Students with Arduino and S4A
    Panagiotis Michalopoulos[1], Sofia Mpania[1], Anthi Karatrantou[2], & Christos Panagiotakopoulos[2]
    1. Introduction

2. Stem Education and Educational Robotics
   3. Arduino
   4. Scratch for Arduino
   5. Methodology
   6. Students' Work and Findings
   7. Discussion and Conclusions

Chapter 8 89
CREATIONS: Developing an Engaging Science Classroom
   Ioannis Alexopoulos[1], Sofoklis Sotiriou[1], Zacharoula Smyrnaiou[2], Menelaos Sotiriou[2], & Franz Bogner[3]
   1. Introduction
   2. Setting of the Study
   3. Initial Results and Discussion

Chapter 9 99
Tangible User Interfaces in Early Year Mathematical Education: An Experimental Study
   Rizos Chaliampalias[1], Anna Chronaki[2], & Achilles Kameas[3]
   1. Introduction
   2. Background
   3. Learning with Tangible Technologies
   4. Material and Methods
   5. Results
   6. Discussion
   7. Conclusions

Chapter 10 109
Utilizing Sphero for A Speed Related STEM Activity in Kindergarten
   Michalis Ioannou[1] & Tharrenos Bratitsis[1]
   1. Introduction
   2. Theoretical Framework
   3. Goal Setting Activities
   4. Discussion

Chapter 11 119
STEM+ARTS=STEAM Skills: Innovation Management and Scratch Programming for Year 4 Students
    Niki Lambropoulos[1] & Ioannis Dimakos[1]
    1. Design Thinking + Hybrid Synergy
    2. Art + Creativity in the Zone of Proximal Flow (ZPF)
    3. Computer Supported Collaborative Creativity + Learning for STEAM
    4. Generative Topics for STEAM
    5. Conclusions + Future Trends

Chapter 12 129
Students' Reasoning with the Pantograph
    Anna Athanasopoulou[1], Michelle Stephan[1], & David Pugalee[1]
    1. Study Design
    2. Findings
    3. Conclusion

Chapter 13 141
Introduction to Physical Computing with Raspberry Pi in a STEM Education Framework
    Nikolaos Balaouras[1], Anthi Karatrantou[2], Georgios Panetas[2], & Christos Panagiotakopoulos[2]
    1. Introduction
    2. STEM Education and Physical Computing
    3. Raspberry Pi
    4. Python for Raspberry Pi
    5. Methodology
    6. Findings and Discussion
    7. Conclusions

Chapter 14 151
Educational Robotics & Children's Attitudes Towards STEM
    Marina Karemfyllaki[1], Anthi Karatrantou[2], & Christos Panagiotakopoulos[2]
    1. Introduction
    2. STEM in Education and Educational Robotics

3. Lego Mindstorms EV3
  4. Children's Attitudes Towards STEM
  5. Research Question
  6. Methodology
  7. Findings and Discussion
  8. Conclusions

Chapter 15 161
From STEM to STEAM and to STREAM Enabled Through Meaningful Critical Reflective Learning
  Vassilios Makrakis[1]
  1. From STEM to STEAM (STEM + The Arts) and Meaningful Learning
  2. The Missing R from STEAM: Shifting to STREAM (STEAM + Reflective Learning)
  3. Concluding Remarks

Chapter 16 173
Bridging STEM (Science, Technology, Engineering & Mathematics) Education with Education for Sustainability
  Nelly Kostoulas-Makrakis[1]
  1. Mainstream STEM Education and Its Contradiction with Efs
  2. Interventions to STEM through RUCAS, CLIMASP and CCSAFS Projects
  3. Conclusion: Lesson Learned

Chapter 17 183
Mechanical Linkages as Links Between Mathematics and Engineering
  Eugenia Koleza[1]
  1. The M Component in STEM Education
  2. Connecting E&M in STEM Education
  3. Our Research: Engineering Mathematics
  4. Methodology
  5. Final Comments

Acknowledgements

The Organizing Committee of these Conferences and the Editors of this Volume would like to thank Ms Vasiliki Psoma and Aspasia Lygda for her skillful work and dedication in the technical editing of this Volume.

Introduction

Eugenia Koleza, Christos Panagiotakopoulos, & Constantine Skordoulis

This volume contains selected papers presented at the Hellenic Conferences "Innovating STEM education HiSTEM 2016 and 2018" organized by the Postgraduate Program "Interdisciplinary Approach on Science, Technology, Engineering and Mathematics in Education STEM Education" (stemeducation.upatras.gr). The first conference was held at the Marasleios Pedagogical Academy of Athens, 16–18 December 2016 and the second at the German School of Athens, 22–24 June 2018. The first eleven papers were presented at the HiSTEM 2016 Conference and the last six papers at the HiSTEM 2018 Conference. These papers were selected after a peer review process from the conferences' submitted papers.

In recent years, there has been a focus on promoting the uptake of STEM subjects in schools. This has been driven by the need to ensure that young people gain the knowledge and skills essential to help them participate in a society in which mathematics, science and technology are increasingly important. Nevertheless, reform efforts, including curriculum development, have treated the STEM subjects mostly in isolation. Recognizing that efforts for education within each individual STEM discipline would encourage a wide range of conservations about different important aspects of teaching and learning, this conference considered the potential benefits and challenges for the integration of various STEM's characteristics into education. In order to prepare students to address the problems of our society, it is necessary to provide them with opportunities to understand these problems through rich, engaging and powerful experiences that integrate the disciplines of STEM.

The conferences provided a platform for dissemination of best practices in teaching and learning STEM in Greece and also inspired and empowered STEM educators to improve teaching quality, to increase engagement in STEM education and career pathways, to connect students with real life industry relevancy and to drive creativity, inquiry-based learning, problem-solving and project-based learning.

The conferences contained four strands:

- STEM Practice: presentations from individuals and teams from K-12 schools who implement STEM programs and learning approaches

- STEM Education Research: presentations from STEM education researchers who examine integrated curriculum challenges in STEM education

- STEM Projects and preliminary results at conception or early practical stages

- STEM Resources included invited organizations and other community partners that have resources available to support STEM education.

The first chapter: "Describing the transition from STEM to STEAM by replicating a historical scientific instrument" by Vasiliki Psoma and Constantine Skordoulis, explores the transition of STEM to STEAM by describing a project designed around the idea of the replication of a historical instrument, for the incorporation of history and philosophy of science in STEM education. A group of students in a primary school is given a set of primary sources, such as historical texts and sketches of historical instruments in order to design their own version of the instrument based on historical principles. Students, while undertaking this project, will be given the opportunity to incorporate other disciplines into their core-content curriculum promoting, in this way, interdisciplinary learning. By injecting creative and innovative thinking into STEM education, with the inclusion of arts, and by crossing boundaries between arts and science, the transition of STEM into STEAM is explored and analyzed.

The second chapter: "The integration of STEM curriculum in the design of a STEM scenario" by John Sdrallis and Eugenia Koleza proposes that education in a STEM environment utilizes the positive elements of integrated teaching but also presents difficulties mainly in planning and implementation. In secondary education, the lack of educators' culture in integrated teaching, the restrictions from the Curriculum regarding the disciplines enclosed in the term STEM, primarily those of science and mathematics and finally the predetermined teaching programme significantly limit any possibility of application. Thus, a key question arises about the extent to which STEM education can be functionally integrated into the existing educational system or constitutes a supplementary action. In the direction of the functional integration, a STEM scenario was planned and implemented in order to record the good practices and difficulties in its application as well as to develop a methodological model that will safeguard the theoretical structure and excellence of a STEM scenario.

The third chapter: "STEMART = STEM + ART: An Erasmus+ STEAM project for K-12 education" by Milan Hausner, Paraskevi Karypidou, Maria Gratiela Popescu, Serkan Altinöz, Laura Bozzo, Olya Mihova, and Petra Mikše is an account of the STEMART Project. The STEMART project focuses on European educational priorities, mainly the support of STEM education. Seven countries (Czech Republic, Greece, Romania, Turkey, Spain, Bulgaria, Slovenia) will organize seven scientific weeks and competitions on chosen topics, to which an international approach will be shown. There are scientific explorations, technical inventions, biological experiments, programming and also construction activities. A permanent sharing of experiences, creation of a scientific ledger and various outputs gives students new horizons and perspectives. As can be clearly seen, a precise evaluation of outputs will create an exceptionally rich scientific and technological experience which can have a strong effect on the future job orientation of students. All products will be published under Creative Commons License. This approach will also involve national bodies, institutions and political representatives in the dissemination of this project. The project will be organized under the auspices of the Czech commission of UNESCO and the representative of European parliament.

The fourth chapter: "Searching for Black Holes. Photometry in our Classrooms" by Ioannis Chiotelis and Maria Theodoropoulou, following the main axis of STEM, presents a module that integrates the use of inquiry-based learning methodology and the integration of ICT tools in schools practices. Is devoted to the implementation of a research-based experiment where students can be involved in the identification of stellar mass black hole candidates and the procedure to "measure" their mass limits. We encourage students to operate the Salsa-J software, perform some photometry measurements and try to spot a Black Hole candidate through images captured by telescopes, available online, or ordered to be taken through a remote telescope. At the end of this module students should know how to identify black hole candidates and how to determine the mass limits of a compact object in a binary star system. Thus, Science, Physics and Astronomy are strongly supported by ICT technology and Mathematics is contributing the data processing and conclusions outcomes.

The fifth chapter: "Developing a Design Framework for UMI Educational Scenarios" by Olga Fragou, Achilleas Kameas, and Ioannis Zaharakis is focused on Ubiquitous learning (u-learning), a new paradigm which is based on ubiquitous computing technology. The most significant role of ubiquitous computing technology in u-learning is to construct a ubiquitous learning environment which enables anyone to learn at any time anyplace. Nonetheless the characteristics of u-learning are still unclear and being debated by the research community. Designing instructional tools that actually promote u-learning experiences is a cumbersome task in the sense of taking into consideration and combining a variety of complex, technological tools and characteristics of u-learning. UMI stands for ubiquitous computing, mobile computing, Internet of Things. This study describes the characteristics and design methodology of a UMI-Sci-Ed Educational Scenario Template as a medium to organize and construct u-learning experiences based in a u-learning environment. It also presents a case study scenario, based on UMI Subject Matter Experts' interaction with the predefined and designed Educational Scenario Components.

The sixth chapter: "New perspectives for Geometry teaching: Mechanical linkages Technology" by Kalliopi Siopi and Eugenia Koleza is a brief presentation of mechanical linkages, and especially of the drawing machines. The focus is on the pantograph, which incorporates mathematical properties and relationships in structure in such a way to allow the implementation one geometrical transformation, such as, symmetry, reflection, translation and homothety. In order to investigate subjects' concepts/ theorems-in-action developed by investigating the structure of the pantograph, and especially the identification of the math concepts and laws incorporated in the machine, the authors selected a pantograph's model and taught homothety to high school students for four hours (early 2016), in the framework of an attempt to incorporate artifacts with the characteristics geometrical machine's in the instruction of Euclidean geometry.

The seventh chapter: "Introducing STEM to Primary Education Students with Arduino and S4A" by Panagiotis Michalopoulos, Sofia Mpania, Anthi Karatrantou, and Christos Panagiotakopoulos is based on STEM education, which is aiming to the development of the scientific interest of students and their capability to solve authentic problems, given emphasis to the connection of Science, Technology, Engineering and Mathematics. They mention that simple applications of automatic control systems and robotic constructions are evolving as basic tools of modern life and these applications provide innovative tools in education as well. In this paper an attempt is made to introduce STEM activities to students of Primary Education supporting them working on simple constructions with Arduino and Scratch for Arduino. Within this framework students were asked to work in groups to design, construct and program their constructions following specially designed worksheets of increasing difficulty with the aim to finally create a Theremin. Analyzing students' work useful information can be derived on how they combined and used in practice knowledge from science, technology and programming and on benefits of educational robotics applications in the frame of STEM education as well.

The eighth chapter: "CREATIONS: Developing an Engaging Science Classroom" by Ioannis Alexopoulos, Sofoklis Sotiriou, Zacharoula Smyrnaiou Menelaos Sotiriou, and Franz Bogner is taking into account the strongly decreased interest of young people in science and mathematics and propose a new creative pedagogic approach in order for this tendency to be reversed, developed in the framework of EU project CREATIONS. The first implementation activities of this approach, addressing both students and teachers, have been already taken place and the initial results of the implementation leads to quite positive conclusions, concerning the motivation and interest in learning science.

The ninth chapter "Tangible User Interfaces in early year mathematical education: An experimental study" by Rizos Chaliampalias, Anna Chronaki, and Achilles Kameas is focused on the use of metaphors in the learning design of tangible user interfaces. Especially for mathematics learning the use of metaphors is paramount for the development of number sense in children. Our research discusses how children can identify multiple correct solutions to numerical problems, and specifically additive structure problems, when conceptual metaphors are being implemented as tangible interfaces, compared to situations where no use of metaphors is made. In order to evaluate the potential for tangible interfaces to support children's numeracy skills, it is important to first identify the possible advantages, as well as limitations, of using conceptual metaphors in this domain.

The tenth chapter: "Utilizing Sphero for a speed related STEM activity in Kindergarten" by Michalis Ioannou and Tharrenos Bratitsis is focused on STEM education in the Kindergarten. STEM education is being gradually adopted by all levels of education and in particular kindergarten lately attracts more attention from the STEM education policy makers because it is believed that children who develop an interest in STEM at a young age are more likely to excel in the future and avoid stereotypes or other obstacles when entering STEM fields in later years. In this paper the design of a teaching activity for approaching the notion of speed in Kindergarten, utilizing the Sphero SPRK robot, is described. The proposed activity is part of a postgraduate dissertation.

The eleventh chapter: "STEM+ARTS=STEAM Skills: Innovation Management and Scratch Programming for Year 4 Students" by Niki Lambropoulos and Ioannis Dimakos mentioned that nowadays, world economic success and well-being require new industries to appear. Technology, Engineering, and Mathematics (STEM) provide tangible solutions towards this direction with STEM introduced in cross-curriculum activities. They propose that when working in Primary Education, STEM is not enough; Arts and Design (STEAM) can introduce fun, surprise, curiosity, teamwork, co- creativity and innovation. STEAM intervention was conducted with Year 4 students at the 6th Primary School of Patras for one year. Initiated within the Flexible Zone hours, STEAM skills development was anchored in Bruner's proposition: anything can be taught at any age if delivered correctly. STEAM advanced concepts were taught for Innovation Management and Programming with Scratch using metaphors and practical examples. Art was approached via Design Thinking within the Zone of Proximal Flow. The creative classroom activities were orchestrated based on Computer Supported Collaborative Creativity and Learning (CSCC/L). The 10 years old students produced innovations and a Scratch game by having fun, innovating and learning together.

The twelfth chapter: "Students' Reasoning with the Pantograph" by Anna Athanasopoulou, Michelle Stephan, and David Pugalee presents the results of a pilot research study they conducted on how the use of a pantograph promotes seventh grade students' mathematical reasoning and argumentation on proportional relationships. Particularly, the analysis of the data indicates students' ability to figure out how a pantograph enlarges or shrinks shapes in a certain scale factor, without having this knowledge before. Also, students found the relation between the scale factor and perimeter and area of rectangles. They combined their knowledge how to calculate perimeter and area of a rectangle and the measurements of length and width of the original rectangle and its image they produced with the pantograph using a different scale factor. In terms of the angles, they estimated that they are preserved and then they verified that they are right angles. Students achieved this goal solving appropriate activities researchers designed and used to guide students' thinking process indirectly.

The thirteenth chapter: "Introduction to Physical Computing with Raspberry Pi in a STEM Education Framework" by Nikolaos Balaouras, Anthi Karatrantou, Georgios Panetas, and Christos Panagiotakopoulos is focused on STEM education, which refers to the scientific approach of solving a problem using tools from Science, Technology, Engineering and Mathematics. This multidisciplinary method allows students to develop a variety of skills needed for their transformation into integrated members of a society where everything is interconnected and continuously is developing technologically. In this paper an attempt was made to introduce students of 9th-grade to the world of STEM education through physical computing using the Raspberry Pi platform and the Python programming language. Students were asked to work in groups to design, construct and program a circuit (system) that simulates the function of traffic lights for visually impaired people. One group of schoolteachers also, attending a STEM postgraduate course, worked with the same tools and the same educational activities in order to discuss their usefulness and applicability in the classroom. Analyzing students' work as well as schoolteachers' thoughts, useful information can be derived on how physical computing can support the aims of STEM education, as well as on the benefit of such educational activities.

The fourteenth chapter: "Educational Robotics & Children's Attitudes Towards STEM" by Marina Karemfyllaki, Anthi Karatrantou, and Christos Panagiotakopoulos is focused on STEM methodology, which in the field of education is commonly understood as an educational approach that includes Science, Technology, Engineering and Mathematics. Educational robotics could be considered as a vehicle for new ways of constructive learning and as a vehicle that leads to new learning paths that are an integral part of STEM education and culture. This research was carried out with the participation of 25 children of the fifth grade of a primary school in a suburb of Athens. The sample divided into groups of 5 members. The children in every group were asked to solve an "authentic" problem with the use of Lego EV3 educational packages. Data were collected with appropriate questionnaires. The results of the data analysis showed that children's attitudes towards STEM-related sciences were more positive and their desire to pursue a career in more than one of the future scientific areas concerning STEM was increased.

The fifteenth chapter: "From STEM to STEAM and to STREAM enabled through meaningful critical reflective learning" by Vassilios Makrakis is based on integrating Science, Technology, Engineering and Mathematics (STEM) subjects, which can be a rewarding process in terms of promoting meaningful learning that is engaging learners among other things in problem-solving, critical thinking and in building real-world connections. However, such a vision is still a quest and STEM or STEAM in its new version has been an area of controversy in terms of outcomes. The question is; how can the content and processes of the four individual subjects become integrated learning areas? How can the integrity of each of these subjects areas be maintained and yet be integrated in a meaningful way? In this paper, the author argue that critical reflection being a driving force towards making STEAM (including Arts) learning more integrated, meaningful and engaging for the students. In this sense, STREAM, integrating R (Reflective learning) is justified and discussed

The sixteenth chapter: "Bridging STEM (Science, Technology, Engineering & Mathematics) Education with Education for Sustainability" by Nelly Kostoulas-Makrakis is focused on STEM, otherwise known as Science, Technology, Engineering and Mathematics, which accounts for most of the skills and knowledge needed in the 21st century. As such, STEM, although originated two decades ago is now making a new impact into education at all levels. Her argument is that STEM, and hence STEM Education discourse has focused more on an instrumental perspective situated in an economic growth ideology that has driven current unsustainable development. In this paper, she presents a critique of the mainstream STEM Education from an Education for Sustainability (EfS) perspective, which argues that STEM needs to shift towards a new paradigm encompassing the principles of EfS. In this way, she outlines the principles of EfS and explore how EfS and STEM Education might be brought together, providing examples of European Commission funded projects that integrate STEM and EfS.

The seventeenth chapter: "Mechanical linkages as links between Mathematics and Engineering" by Eugenia Koleza contains a research that aims at establishing communication bridges between two apparently disconnected academic communities, the mathematicians' and the engineers', through a common tool: simple and mathematical machines. Mechanical devises may be used as a context for both reinforcing students' spatial abilities and strengthen engineers' construction skills. The main goal is to show the importance of introducing a new register of a concept in mathematics course the mechanical register in order to improve the students' understanding and learning. This paper reports part of a study investigating the use of a specific mathematical machine, the pantograph for establishing an interdisciplinary culture of apprehending science and mathematical concepts and engineering design.

This volume, including the abovementioned selected contributions from the HiSTEM 2016 and 2018 conferences, envisages that the readers will be informed on how to: make ICT real world relevant but still cover curricular outcomes, understand the different types of projects they can implement, design a project-based Math and Science learning unit, run assessments in a project-based learning setting, explore best practices for mentoring and training STEM students to develop the knowledge and skills necessary to connect their learning to real-world problems and identify institutional practices for integrating the values and traditions of STEM disciplines with other campus domains to fully capture the distinct characteristics and comprehensive nature of scientific solutions. Readers' feedback on the aims and scope of this publication will be very much appreciated by the editors of this volume.

Chapter 1

Describing the Transition from STEM to STEAM by Replicating a Historical Scientific Instrument

Vasiliki Psoma[1] & Constantine Skordoulis[1]

[1]*Department of Primary Education, National and Kapodistrian University of Athens, Greece*

Abstract: *This paper explores the transition of STEM to STEAM by describing a project designed around the idea of the replication of a historical instrument, for the incorporation of history and philosophy of science in STEM education. A group of students in a primary school is given a set of primary sources, such as historical texts and sketches of historical instruments in order to design their own version of the instrument based on historical principles. Students, while undertaking this project, will be given the opportunity to incorporate other disciplines into their core-content curriculum promoting, in this way, interdisciplinary learning. By injecting creative and innovative thinking into STEM education, with the inclusion of arts, and by crossing boundaries between arts and science, the transition of STEM into STEAM is explored and analyzed.*

Keywords: *STEM, STEAM, Interdisciplinary, Integrated Curriculum, historical replication, project method*

1. Introduction

It is widely accepted that students, through historical experiments, can gain an important insight into the way scientists work, which contributes to a more meaningful understanding of the history of science [see Holubová (2014) and references therein]. Slykhuis et al. (2015) have recently substantiated that the implementation of historical reconstructions can enhance learners' STEM experiences by combining all aspects of STEM and by meeting curricula standards across science, mathematics, engineering, and technology. By breaking down the traditional barriers between STEM disciplines, interdisciplinary learning can be promoted in an authentic manner. In our paper, the inclusion of the discipline of history of science in an interdisciplinary program is proposed as has been previously by Viterbo (2007).

Land (2013) recognizes the value of arts integration into the STEM curriculum, which transforms STEM into STEAM, stressing that this opens up multiple pathways for individual meaning-making and self-motivation. An effective collaboration among teachers on arts integrated project-based learning will promote, not only understanding of the content, but literacy in general. In the same vein, Milkova et al. (2013) argue that the integration of the two can nurture students' higher-order thinking skills and develop skills inherent in such thought, such as analyzing, synthesizing and evaluating as well as stimulate collaborative and creative thinking.

Primary education offers fertile ground for the implementation of integrative approaches to STEM education, since a teacher at the primary school level is responsible for the teaching of a variety of modules (Quang et al., 2015; Sanders, 2009). Even though the concept of STEM education is not new, a considerable number of teachers appears unaware as of how to operationalize STEM education (Kelley & Knowles, 2016) and/or they follow a Silo STEM instruction (Raman, 2011; Wang et al., 2011).

Furthermore, there are only few reports available addressing the issue of historical reconstruction in the primary school. Even though some action has been taken to incorporate the historical, social and philosophical aspects of science into the curriculum, the laboratory conditions that could synthesize an authentic experience continue to be "absent" (Höttecke, 2000).

In STEM education, interest is shifted to the design part-through the emphasis on engineering (Bequette & Bequette, 2012; Wang et al., 2011) and scientific inquiry (Kennedy & Odell, 2014). From a pedagogical perspective, art and engineering education are related to problem-based learning (PBL), since both are looking for visual solutions through the design process (Bequette & Bequette, 2012).

This paper explores the constitution of STEAM by describing a project designed around the idea of the replication of a historical instrument in an attempt to incorporate of history and philosophy of science in STEM education. A group of students in a primary school is given a set of primary sources, such as historical texts and sketches of historical instruments in order to design their own version of the instrument (camera obscura) based on historical principles. Students, while undertaking this project, will be given the opportunity to incorporate other disciplines into their core-content curriculum promoting, in this way, interdisciplinary learning.

2. Theoretical Framework

The use of history and philosophy in science teaching has been suggested by many studies (Höttecke et al., 2012), which have described the pedagogical advantages from the utilization of history of science-to-science education, such as encouraging student motivation while contributing to an effective curriculum project on science (Brush, 1989).

It has to be noted that there is considerable difficulty on behalf of the students to apply their knowledge to different disciplines (Figliano, 2007), something that can be partly attributed to teachers' weakness in identifying logical connections between different subjects areas (Wineburg & Grossman, 2000). This constitutes a strong argument in favor of interdisciplinary approach to learning, otherwise known as STEM education (Tsupros et al., 2009).

STEM highlights the implementation of higher-level skills for students and aims at fostering inquiring minds, with an ultimate goal to develop interdisciplinary thinkers able to generate new ideas by making connections between seemingly disparate concepts (Figliano, 2007).

Although it has been repeatedly stated that a subject cannot be studied in isolation from the arts, humanities, and social studies (Sanders, 2009), STEM subjects appear to be disconnected from arts, creativity, design (Hoachlander & Yanofsky, 2011). While recruiting and retention efforts are deemed imperative to make sure that more and more students are leaning towards STEM fields, arts are being squeezed out of schools and marginalized in the educational system (Wynn & Harris, 2013), as they are no longer considered indispensable (LaMore et al., 2013).

Taylor (2008) argues that the inclusion of the arts can have a high pedagogical value, while further stating that interdisciplinary work can broaden students' knowledge of history and different cultures. Integrating arts into core content areas, apart from taking into account all different learning modalities, it facilitates students to examine a concept from various perspectives, since well-rounded problems related to art have more than one single answer.

The so-called STEAM movement's prime concerns are about the promotion of skills, such as creativity, experimentation, innovation and problem solving, which would entail the empowerment of science learning (Cultural Learning Alliance, 2014).

Learning theories emphasize the situatedness of learning, by positing that learning is situated within an authentic activity and context. When setting the integration of STEM content under consideration, engineering design context can be the situated context for learning, given that the engineering element focuses on the generation of alternative solutions to real world problems instead of identifying a unique, optimal solution.

Project-Based Learning by incorporating engineering design principles within the curriculum creates the context from which genuine experiences for students can arise and provides the foundation on which students can construct concepts related to science, technology, engineering and mathematics, with the support of art, towards the unification of fragmented pieces of knowledge, in order to make real-world connections (Capraro & Slough, 2013). This position suggests STEM Project-Based Learning as the base of integrating STEM into curriculum.

2.1. Aims and Scope

Few are the publications in the literature dealing with the issue of historical replication in primary school, exploring the contribution of the historical component to a STEAM educational approach to be placed in the broader context of an integrated curriculum.

Another reason advocating the selection of the present topic is related to the fact that the history of science in education is not among students' favorite subjects, and information about the life of known scientists is of low interest in primary schools (Dvořák 2008, as cited in Holubova, 2014, 163).

In the wider context of utilizing the history and philosophy of science, this proposal deals with the historical reconstruction of a historical instrument; camera obscura. The stages of development that camera obscura has gone through and evolved into a dark box with a lens and mirror, as well as the number of scientific fields it has been linked to over the centuries, make it an ideal topic of further study, creating a fruitful ground for a multifaceted approach to education.

Its relation to optics and to vision theories that students elaborate in these classes (Wade & Finger, 2001) contributed positively to the selection of this historical instrument as the topic of our research proposal. Besides that, the art of design in education is one of the fundamental components of STEAM education, and this instrument is largely related to the history of art (Lefèvre, 2007).

2.2. Description of Related Activities

Camera obscura (pin-hole camera) and lens images fall within the topic of optics introduced in the National Curriculum of the primary school in Greece. Considering the fact that the topic of optics appears in the last grade of the primary school in Greece, students are not expected to achieve a deep understanding of core cognitive aspects (use of precise mathematical formulas), but to develop key skills, such as organizational, creative, and communication skills (Bybee et al., 2008) and acknowledge different aspects of the nature of science (NoS).

A complementary replication of camera obscura will be attempted, in that each cohort will work on its version of the instrument by modifying -at will- the reconstruction process suggested by primary textual sources and by utilizing alternative materials.

By being actively involved in the design process in a variety of ways (Engineering) by taking measurements and by attempting to change the dimensions of the instrument (Mathematics), and by presenting, in the end, a practical application of this process (Technology), students will address operational principles crossing curricular lines. In relation to other optical instruments, what differentiates and makes camera obscura worthwhile to study further is the fact that its layout and operation resemble the structure and function of the eye, contributing to the representation of the visual process (Lefèvre, 2007). This way, additional links to other subjects, such as biology, are reinforced.

With regard to the arts integration, this instrument, after its completion, can be used in various ways by students, as, for example, in creating drawings of landscapes and/or observing other persons. Students can obtain an understanding of the potential of a scientific device in the arts, since they will use it to support their drawing. Additionally, students' prior knowledge of the light properties can be used to help them ensuring that their images are as high-quality as possible, gaining numerous perspectives on the history of photography. After that, pupils will be able to take individual or group pinhole portraits and even set up their own darkroom. Through this process, learners will learn to detect the points in which a camera obscura and a modern camera differ, gaining a deeper understanding of the operating and technical principles that govern them, and getting themselves into the world of photography.

Moreover, students become familiar with the works of famous painters, like Leonardo da Vinci, who studied the laws of geometric perspective and used camera obscura in his attempt to adopt a system of perspective design in his paintings (Farago, 1991), and Vermeer, who seems to have used camera obscura as an aid in creating his paintings (Lefèvre, 2007).

3. Educational Approach: EDP and Project-Based Learning (PBL)

The incorporation of authentic tasks, like constructing an artifact, is considered to be among the advantages of the integration of STEM and PBL (Scheurich & Huggins, 2008). This is the reason why a Project-Based learning approach is adopted for this study.

This project will be conducted in a private school, which offers the International Baccalaureate (IB) program for students. This activity will take place once a week lasting about 45 minutes. The present project will take 24 weeks to complete. Considering the fact that it is an IB school and that our study sample will consist of children between the age group of 11-12 years, we expect that pupils at that age will be acquainted to a certain extent with the concept of Project-Based Learning (PBL) which is the cornerstone of the IB's educational philosophy.

For the project-based learning approach students will use Engineering Design Process (EDP) and will work in groups for the implementation of the design project. EDP has been described as a decision-making process usually being iterative in nature with one of its true values being the development of critical thinking and meta-cognitive skills, since optimal solutions are designed by students with implications for their daily lives (Mangold & Robinson, 2013). Specifically, science and mathematics concepts are applied to engineering projects, addressing also to students who do not deal with core science and math concepts specified by the curriculum. The culminating event of the whole engineering design procedure would be the delivery of an engineering presentation, including a discussion upon the societal impact of the solution(s).

Such an engineering STEAM education plan, which uses project/problem-based (PBL) learning as its vehicle, enhances skills, such as collaboration, communication, cooperation, creativity, critical thinking, and problem-solving, which comprise the essence of STEAM education (STEAM Team Committee, 2013).

In Table 1, we can see the phases of the EDP (Massachusetts Department of Education, 2001) which will be applied to the project.

Table 1: Steps of the Engineering Design Process

Engineering Design Process
Identify the need or problem
Research the need of problem
Develop possible solutions
Select the best possible solution(s)
Construct a prototype
Test and evaluate the solution(s)
Communicate the solution(s)
Redesign

The steps of the EDP process will be presented to the classroom before the project begins. It is expected that by providing students with the clear steps of the EDP and by allowing them to follow a flow chart method, group decision-making about which step should be followed next will be more efficient and effective (Mangold & Robinson, 2013). Moreover, linking students' curriculum to engineering and hands-on projects will cause greater student engagement for the subjects falling under this study.

Studying excerpts from primary sources, at the beginning of the project, will shed light on socio-cultural aspects of camera evolution over the years, introducing students to the history of science. With visual aid tools being offered to students, such as sketches of historical instruments, the brainstorming phase will take place-phase helpful in identifying prospective alternatives and in stimulating the building of ideas and brainstorming techniques, such as brain writing and mind mapping will be undertaken.

After that, each group will be asked to draft a sketch of the instrument, including details about its dimensions and scaling, as well as the materials it will be made from. Engineering drawing concepts will make their appearance at this point, guiding students in the drafting process. In addition to this, the identification of the purpose of each design component is deemed necessary, encouraging students to determine its exact function.

Before finalizing their design decisions and entering the final design stage, students are expected to test and modify their replica according to certain criteria. The horizontal use of technology (tablets, 3D printer) in the given context will have both practical and aesthetic dimensions, since it will help in producing aesthetic forms.

It is expected that students, during the process, will investigate the properties and spatial features of the individual components of their constructions and will measure their dimensions, considering the function of each one of them. This way, spatial awareness and geometrical drawing skills will be developed. The ability to calculate area, perimeter and volume of these components is crucial to construction projects.

Assessment of the project will be partly based on a group presentation of the final product and a project report providing students' critical reflection. Although in the majority of engineering programs, learning assessment places emphasis on disciplinary content, a same focus is required for assessing personal and interpersonal skills, as well as design skills (Núñez et al., 2013). Hence, the "success" of each team will not depend exclusively on well-rounded, standards-based performance but also on the way that each student will follow the engineering design process steps. Interviews with the students and classroom observations will also take place.

For a STEM project, information stemming from both formative and summative assessment outlines a number of guidelines regarding the identification of errors, the modification of the process and evidence describing whether the project worked the way it was supposed to (Fragoulis & Tsiplakides, 2009).

4. Discussion

This paper describes a historical instrument replication project, which aims at incorporating the historical dimension into the traditional STEM curricular areas, and advocates for the inclusion of history of science into an interdisciplinary course, adopting a STEAM educational approach in the direction of an integrated curriculum, and investigating the application of this integrative approach in a situated learning context in Greece.

In terms of the aforementioned ways, learners should identify connections between different content areas and apply their knowledge in various disciplines, such as Science, Technology, Engineering and Mathematics, with the inclusion of Arts, recognizing their interdependence, thus an interdisciplinary approach to learning is adopted. In the forefront comes the development of skills, such as critical-thinking, innovative thinking, collaborative, and problem-solving skills.

With the proposed intervention, we expect that students will be able to design and build a replica, using fundamental engineering principles. At the conclusion of the project, students should be able to summarize and evaluate their ideas and insights, and communicate them to the broad audience by means of presentations and reports and by using technology as a learning tool.

Finally, students, through the critical analysis of some indicative passages coming from original historical sources, with information on the social and cultural context of each time period, are expected to demonstrate an insight concerning their understanding of Nature of Science (NOS). Regarding the objectives related to the understanding of NOS, students are expected to be able to create connections between seemingly distinct areas, such as Science and the Arts, as well as to recognize the interrelationship of Science, Technology and Society (STS) and be able to argue about their mutual dependence (Teichman, 1991).

REFERENCES

Bequette, W. J., & Bequette, M. B. (2012). A place for art and design education in the STEM conversation. *Art Education, 65*(2), 40–47. https://doi.org/10.1080/00043125.2012.11519167

Brush, G. S. (1989). History of Science and Science Education. *Interchange, 20*(2), 60–71.

Bybee, W. R., Powell, J. C., & Trowbridge, L. W. (2008). *Teaching secondary school science: Strategies for developing scientific literacy* (9th ed.). Prentice Hall.

Capraro, M. R., & Scott, W. S. (2013). Why PBL? Why STEM? Why now? An introduction to STEM project-based learning: An integrated science, technology, engineering, and mathematics (STEM) approach. In R. M. Capraro, M. M. Capraro, & J. Morgan (Eds.), *Project-based learning: An integrated science, technology, engineering, and mathematics (STEM) approach* (2nd ed., pp. 1–6). Sense.

Cultural Learning Alliance (2014). STEM+ARTS = STEAM. https://www.culturallearningalliance.org.uk/images/uploads/STEAM_report.pdf

Farago, J. C. (1991). Leonardo's Colour and Chiaroscuro Reconsidered: The Visual Force of Painted Images. *Art Bulletin, 73*(1), 63–88.

Figliano, F. (2007). *Strategies for Integrating STEM Content: A Pilot Case Study* (Unpublished master's thesis). Virginia Polytechnic Institute and State University.

Fragoulis, I., & Tsiplakides, I. (2009). Project-Based Learning in the Teaching of English as a Foreign Language in Greek Primary Schools: From Theory to Practice. *English Language Teaching, 2*(3), 113–119. https://doi.org/10.5539/elt.v2n3p113

Hoachlander, G., & Yanofsky, D. (2011). Making STEM real: by infusing core academics with rigorous real-world work, linked learning pathways prepare students for both college and career. *Educational Leadership, 68*(3), 60–65.

Holubová, R. (2014). Historical Experiments in Physics Teaching. *US-China Education Review, 4*(3), 163–172.

Höttecke, D. (2000). How and what can we learn from replicating historical experiments? A case study. *Science and Education, 9*(4), 343–362.

Höttecke, D., Henke, A., & Riess, F. (2012). Implementing history and philosophy in science teaching: Strategies, methods, results and experiences from the European HIPST project. *Science & Education, 21*(9), 1233–1261. https://doi.org/10.1007/s11191-010-9330-3

Kelley, R. T., & Knowles, G. J. (2016). A conceptual framework for integrated STEM education. *International Journal of STEM Education, 3*(1), 1–11. https://doi.org/10.1186/s40594-016-0046-z

Kennedy, T., & Odell, M. (2014). Engaging students in STEM education. *Science Education International, 25*(3), 246–258.

LaMore, R., Root-Bernstein, R., Root-Bernstein, M., Schweitzer, J. H., Lawton, J. L., Roraback, E., Peruski, A., VanDyke, M., & Fernandez, L. (2013). Arts and crafts: Critical to economic innovation. *Economic Development Quarterly, 27*(3), 221–229. https://doi.org/10.1177/0891242413486186

Land, H. M. (2013). Full STEAM ahead: The benefits of integrating the arts into STEM. *Procedia Computer Science, 20*, 547–552. https://doi.org/10.1016/j.procs.2013.09.317

Lefèvre, W. (2007). *Inside the Camera Obscura Optics and Art under the Spell of the Projected Image.* Max-Planck-Institut für Wissenschaftsgeschichte.

Mangold, J., & Robinson, S. (2013, June 23–26). *The engineering design process as a problem solving and learning tool in K-12 classrooms* [Paper presentation]. 120th American Society for Engineering Education (ASEE): Annual Conference & Exposition, Atlanta, GA, USA. https://www.asee.org/file_server/papers/attachment/file/0003/4031/7971.pdf

Massachusetts Department of Education (2001). *Massachusetts science and technology/engineering curriculum framework.* Massachusetts Department of Education.

Milkova, L., Crossman, C., Wiles, S., & Allen, T. (2013). Engagement and skill development in biology students through analysis of art. *Cell Biology Education, 12*(4), 687–700. https://doi.org/10.1187/cbe.12-08-0114

Núñez, J., Lascano, S., & Esparragoza, I. (2013, August 14–16). A project-based learning approach for a first-year engineering course. In M. M. Larrondo Petrie, H. Alvarez, I. E. Esparragoza, & C. Rodriguez Arroyave (Eds.), *Innovation in Engineering, Technology and Education for Competitiveness and Prosperity: Proceedings of the 11th Latin American and Caribbean Conference for Engineering and Technology.* LACCEI Inc. http://www.laccei.org/LACCEI2013-Cancun/RefereedPapers/RP247.pdf

Quang, L. X., Hoang, L. H., Chuan, V. D., Nam, N. H., Anh, N. T., & Nhung, V. T. (2015). Integrated Science, Technology, Engineering and Mathematics (STEM) Education through Active Experience of Designing Technical Toys in Vietnamese Schools. *British Journal of Education, Society & Behavioural Science, 11*(2), 1–12. https://doi.org/10.9734/BJESBS/2015/19429

Raman, N. (2011). Digital Provide: Education Beyond Borders. In C. Wankel & J. S. Law (Eds.), *Streaming Media Delivery in Higher Education: Methods and Outcomes* (pp. 1–461). IGI Global.

Sanders, M. (2009). STEM, STEM education, STEM mania. *Technology Teacher, 68*(4), 20–26.

Scheurich, J. J., & Huggins, K. (2008). Preface. In R. M. Capraro & S. W. Slough (Eds.), *Project based learning: An integrated science technology engineering and mathematics approach* (pp. vii–x). Sense.

Slykhuis, A. D., Martin-Hansen, L., Thomas, C. D., & Barbatom, S. (2015). Teaching STEM through historical reconstructions: The future lies in the past. *Contemporary Issues in Technology and Teacher Education, 15*(3), 255–264.

STEAM Team Committee (2013). *Planning for the Plan.* http://www.edleader21.com/info/CCPS-STEAM-whitepaper.pdf

Taylor, J. A. (2008). From the Stage to the Classroom: The Performing Arts and Social Studies. *History Teacher, 41*(2), 235–248.

Teichman, J. (1991). History and historical experiments in physics education with special regard to astronomy. *Physics Education, 26*(1), 46–51.

Tsupros, N., Kohler, R., & Hallinen, J. (2009). *STEM education: A project to identify the missing components.* Intermediate Unit 1: Center for STEM Education and Leonard Gelfand Center for Service Learning and Outreach, Carnegie Mellon University, Pennsylvania.

Viterbo, P. (2007). History of Science as Interdisciplinary Education in American Colleges: Its Origins, Advantages, and Pitfalls. *History of Science, 3*(2), 1–19.

Wade, J. N., & Finger, S. (2001). The eye as an optical instrument: From camera obscura to Helmholtz's perspective. *Perception, 30*(10), 1157–1177. https://doi.org/10.1068/p3210

Wang, H.-H., Moore, T. J., Roehrig, G. H., & Park, M. S. (2011). STEM integration: Teacher perceptions and practice. *Journal of Pre-College Engineering Education Research, 1*(2), 1–13. https://doi.org/10.5703/1288284314636

Wineburg, S. S., & Grossman, P. (2000). *Interdisciplinary curriculum: Challenges to implementation.* Teachers College Press.

Wynn, T., & Harris, J. (2013). Toward a STEM + arts curriculum: Creating the teacher team. *Art Education, 65*(5), 42–47. https://doi.org/10.1080/00043125.2012.11519191

Chapter 2

The Integration of STEM Curriculum in the Design of A STEM Scenario

John Sdrallis[1] & Eugenia Koleza[1]

[1]*Laboratory of Research in Mathematics Teaching, Department of Primary Education of the University of Patras*

Abstract: *Education in a STEM environment utilizes the positive elements of integrated teaching but also presents difficulties mainly in planning and implementation. In secondary education, the lack of educators' culture in integrated teaching, the restrictions from the Curriculum regarding the disciplines enclosed in the term STEM, primarily those of science and mathematics and finally the predetermined teaching programme significantly limit any possibility of application. Thus, a key question arises about the extent to which STEM education can be functionally integrated into the existing educational system or constitutes a supplementary action. In the direction of the functional integration, a STEM scenario was planned and implemented in order to record the good practices and difficulties in its application as well as to develop a methodological model that will safeguard the theoretical structure and excellence of a STEM scenario.*

Keywords: *integrated STEM education, criteria of good STEM practices, biodiesel, life-cycle assessment, iLUC indirect Land Use Change*

1. Introduction

The sectors of the whole spectrum of the human activity are becoming increasingly interrelated and interdependent. Communication networks of exchanging information throughout the world create new forms of cooperation and transform the nature of work and knowledge. New areas of research are developed in order to promote human knowledge so as to correspond to the challenges of this changing world with insight and novelty. These include areas emerging from the combination of multiple fields such as science, technology, engineering, arts etc.

Hence, this new reality affects education as well. Today's students are invited to confront a unique range of social, scientific, financial, cultural, environmental, political and technological problems. In order to respond to these effectively, they have to possess interdisciplinary skills that relate to investigating, managing information, cooperating, thinking creatively and critically and applying technologies. Students should be acquainted with new methods and forms of analysis, interpretation, synthesis and evaluation which will allow them to acquire the skills of applying contemporary systems of thinking and planning, to conduct scientific inquiry, to simulate ideas by constructing and using models and prototypes, to produce new products and to apply solutions, exceeding skills and knowledge of established fields (Ontario Ministry of Education, 2002). Education in a STEM setting as an integrated teaching approach of interdisciplinary type, aims at the acquisition of such skills.

Education in a STEM setting is a teaching approach that integrates the content and skills of natural science, technology, engineering and mathematics. The criteria of good practices that define specific behaviours in combination with the content of the STEM disciplines constitute what is expected from an adept STEM student. In summary, these criteria are: a) the in-depth learning and the ability to apply the content of natural science, technology, mathematics and engineering, b) the combined use of the content of the STEM disciplines in order for the student to be capable to answer complex questions and solve complicated problems, c) the capacity to interpret and manage information related to natural science, technology, engineering and mathematics, d) the ability to conduct research, e) the development of correct reasoning, f) the cooperation among groups in the context of STEM and g) the application of technology in a strategic manner (Maryland State STEM Standards of Practice, 2012).

The "science" and "mathematics" components are to a great extent more familiar in regard to their content and learning objectives compared to the "technology" and "engineering" ones which are less known. Technology aims at the creation of product objectives and the solution of problems whereas science at the description and interpretation of world phenomena (Norström, 2013). Even though both are different scientific fields of varied methods and results, a binary relation between them is developed in two levels. On a first level, technology derives from science.

Among students and not only there is the stereotypical belief that science precedes technology, a fact that in many cases is confirmed. Science indeed motivates and leads the technological development. Semiconductors, microelectronics, lasers, nanotechnology emerged from scientific discovery. There are, however, several occasions that support the opposite. People knew about copper processing since the last millennium of pre-Christian times, long before the definition of the terms of oxidation and reduction and England gained industrial power in the 18th century, immediately after the invention of steam engine and well before the understanding of the theoretical framework for the production of energy through thermal machines.

On a second level that of education technology and its application encourage students to discover scientific concepts, whose utility and value is hard to grasp as they seem abstract and fragmentary in relation to the real world. This motive extends beyond technology and includes the whole socio-cultural framework which influences learning greatly since within it, knowledge includes the social interactions in which it is classified (McCormick, 2006). The creation of such motives constitutes the strongest argument of the supporters of the STEM integration model, who consider science and technology as the appropriate vehicle for teaching problem solving skills, emphasizing methods and practices that scientists and engineers use in their respective STEM professions.

Nonetheless, in secondary education, the curriculum of most counties is completely distinct without demonstrating the structural relations among subjects and without leaving space for integrated teaching. Since STEM education is not intended to comprise a new autonomous thematic area taught by specialized educators, its application is to be fulfilled by the current curricula and by the teachers teaching the subjects enclosed in STEM acronym (Sanders, 2008). Therefore, teachers of STEM subjects should possess the appropriate tools in order to prepare and employ integrated teaching surmounting the adversities and restrictions caused by the structure and the function of school units.

The authors of the Next Generation Science Standards (NGSS) framework claim that efforts should be made to reinforce the scientific training of teachers in all educational levels without changing the existing programmes of studies. They also underline the value of skilful "weaving" of the common practices of science, engineering, technology and mathematics around the core of important scientific concepts. They contend that this project constitutes the main challenge since it is impossible to coordinate the programmes of studies (Barakos et al., 2012).

The present study explores the way this coordination-i.e., the alignment of a STEM scenario with the curriculum aims and objectives may be accomplished-, while employing STEM type activities.

2. Teaching Natural Science in STEM-Based Scenarios

The pursuit of creating students with STEM skills to take up STEM related studies and professions are at the core of this process (STEM college and career ready students). This can be achieved with an integrated curriculum of mathematics, science, technology and engineering. By teaching intervention, students need to meet their goals and aims of curricula which in our case are national and identical for all schools. Their achievement is fulfilled in a teaching environment of collaborative kind which emphasizes practices and methods through which students construct/discover knowledge, addressing real world problems, constructing explanatory figures, models, prototypes as well as products that they evaluate and redesign meeting the criteria of good practices of STEM (Learning Environment and STEM standards of Practice).

From the study of literature, it is evident that there is not a clear and predetermined methodological standard neither for the agenda nor the planning and implementation of STEM scenarios. A review of the definitions of STEM demonstrates that the main idea is summarized in the common connections across the STEM subjects, in the creation of motives for a STEM career, in the interdisciplinary approach and primarily in the students' exposure to authentic experiences in strict application of the content for the solution of real problems (Burrows et al., 2014).

The solution of real-world problems that are open-end and ill-structured, in a Problem-Based Learning teaching context generates an appropriate field of acquiring STEM skills. The interdisciplinary nature of STEM scenarios improves the students' performance according to many. However, in order to investigate the effects and impact of such teaching approaches on the student's performance, further research is required as the current research is in an initial level (Tseng et al., 2013). The integrated teaching in a Problem-Based Learning teaching context provides students with, inter alia, the opportunity to reflect on the acquired knowledge and to comprehend the way in which learning is used for the solution of problems (Hmelo-Silver, 2004). The record of this reflection as well as the students' arguments during the process of searching for solution are favoured in this teaching environment, offering the possibility of inquiry and evaluation of the impact of teaching.

2.1. The Topic Choice

Starting from the basic idea of climate change and seeking for solutions for inverting it, we went across the curricula of chemistry and biology horizontally (i.e., in the second class of senior high school) and through their aims and objectives, we searched for an authentic real-world problem to design and apply a STEM based teaching. The teaching is of Problem-Based Learning with activities that require the application of science and technology methods, for example the scientific inquiry (Inquiry Based Scientific Education, IBSE) and the engineering design process (EDP).

According to the previous statements, the selection of topic was based on three conditions: the importance of the subject, the possibility of interdisciplinary approach and the possibility of bringing out the desired skills. The topic of choice that also meets the previous criteria was that of "biodiesel". Within the framework of investigating this topic, students will confront an authentic real-world problem and they will acquire knowledge and problem-solving skills through the EDP and scientific inquiry (Laboy-Rush, 2011).

The climate change is a problem that often generates controversy. Most students of secondary education realize the problems brought about by climate change (pollution, depletion of natural resources, etc.) but they do not understand their size and extent. Climate change is attributed to a great extent to the greenhouse effect and increasing emissions of CO2 which will constitute the connection with fuels and renewable sources of energy whose product is biodiesel.

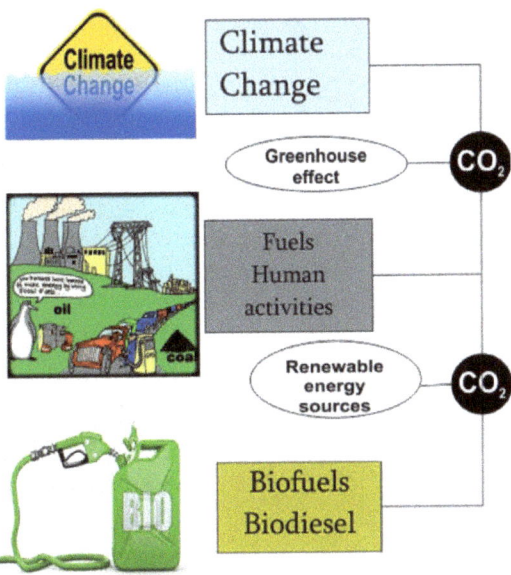

Figure 1: The basic idea of the STEM scenario

The production of biofuels and more specifically of biodiesel can be used in full alignment with the curricula of science and particularly those of chemistry and biology of the second class of senior high schools. The alignment is achieved, as shown later, through the big ideas and the curriculum aims of those subjects as they are defined in international literature. In light of this, the teaching aims are selected for the completion of which, specific STEM activities are planned. As students develop the knowledge base of the components of science through scaffolding, they will apply knowledge in new fields as that of renewable sources of energy, incorporating appropriately Mathematics, Technology and Engineering (Burrows et al., 2014).

3. "Integrating" The Curricula

The implementation of STEM scenarios is accomplished by one or two teachers, in one or two classes, in different teaching periods (Moore & Smith, 2014) and includes cooperative activities. It is a difficult task both in terms of planning as well as in terms of implementation and classroom management (Smith et al.,

2009). Regarding the implementation and management, the difficulties are due to the fact that the teachers who will implement the STEM scenarios are not necessarily the same people who will have designed the activities and they are not often adequately trained in such planning and have insufficient experience regarding cooperation with other teachers. A significant difficulty of STEM scenarios implementation in schools is that the connection of cognitive objectives of the scenarios and those of the curriculum of each one of the STEM components is not clear. There are two ways of STEM integration. The first one concerns the realization of an off-curriculum project with the participation of many teachers. The students together as a group participate in activities that are supervised and assisted by teachers from whom the students seek advice and support. The second way is a complete integration in which a teacher follows a thematic project of many teaching periods, as it usually occurs in primary education. If this is implemented in secondary education, the teacher carrying it out is required to possess diverse skills beyond those of the specific subject so as to achieve the appropriate format (Banks & Barlex, 2014). In terms of the aforementioned ways, it should be noted that in Greece, STEM education attempts follow the first type of integration. In particular, in private schools, summer STEM camps or STEM activities outside the curriculum have been organized during the school year or at weekends. Additionally, private entities organize festivals including STEM activities in order to bring students closer to science and technology. Contrary to previous attempts, the present study describes a procedure of planning a STEM scenario which will be implemented and evaluated during a school year by one teacher in the syllabus of the senior high school.

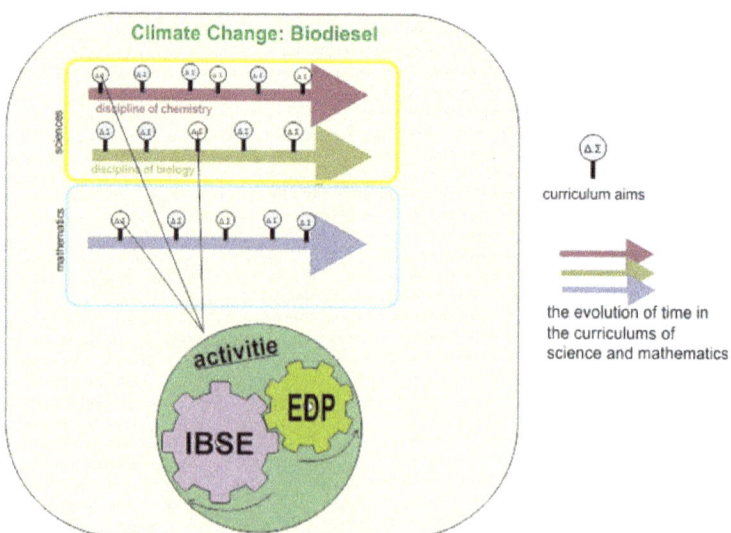

Figure 2: The basic idea as a criterion for the selection of teaching aims in creating activities

In terms of our proposal, we planned the intervention in a way that the learning objectives of the integrated subjects and especially those of science, engineering and mathematics will be considered for the solution of a real-world problem, namely "biodiesel", as a response to the climate change phenomenon.

The planning of activities follows the "backward design process" methodology, which is adapted to the integration requirements (Wiggins & McTighe, 2005). The steps of this plan are: i) the identification of the desired outcomes ii) the determination of the expected end results so as to confirm that the initial objectives were met iii) the design of the teaching intervention which will achieve the desired results. When the design follows these steps, then the content, the assessment and the teaching strategy are coherent (Wiggins & McTighe, 2005). In the case of STEM scenarios in which the design must take into account integration, it is necessary to add one more steps in response to the question: can the goal and the learning objectives be organised in a meaningful way? In the following table, the steps and the corresponding questions for the planning of an integrated teaching are described (Drake & Burns, 2004).

Table 1: Expanding the Principles of backward curriculum design for the planning of an integrated teaching

	Backward Design Process	Questions
1.	The identification of the desired outcomes	• What is important and what is required for understanding? • What kind of impact do we want the teaching to have on students?
2.	Review of the standards in order to determine the use within an interdisciplinary framework	• Can the standards be organised in a meaningful way?
3.	The determination of the evidence	• What is the evidence of understanding?
4.	The design of teaching intervention that will bring the desired outcomes	• What are the learning experiences that promote understanding and lead to the desired results?

4. Biodiesel": A Teaching Proposal

"Biodiesel" is in the planning and implementing phase (work in progress). The activities are designed based on the initial draft that includes three areas-phases. These are determined by common characteristics such as the creation of prototypes, the application of technology, the mathematical processing, the logical reasoning and the combination of those.

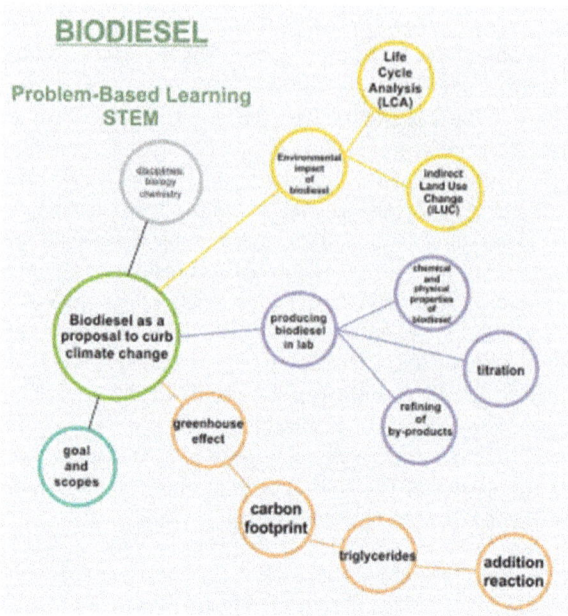

Figure 3: The initial draft which includes three areas-phases

The common characteristic of the first phase (activities in orange in picture 4) is the definition of the real-world problem that the students need to solve. As the climate change and its connection with human activities is the main idea, students will be actively involved in teaching, connecting the main idea with biodiesel. After a series of activities, they will formulate and search for answers to questions like: How does the greenhouse effect connect with human activities? What is the carbon footprint and how can it be diminished? What proportion corresponds to transport? What are the renewable sources of energy? Why is biodiesel part of renewable sources of energy, what are its advantages and disadvantages compared to fossil fuels? What are the appropriate raw materials for its production and how these affect the final product? The aim of this first phase is to justify the decision of the leaders of developed and developing countries to replace fossil fuels used today in the transport sector (an important percentage of the total fossil fuel consumption reaches almost 28%) with biofuels, in order to confine the CO_2 emissions that mainly contribute to the climate change through the greenhouse effect.

The Engineering Design a key component in STEM teaching applies to "biodiesel" in a second phase, as students in the Science laboratory will produce biodiesel through a succession of simple steps. The materials that they will use for its production are cheap and easy to find. Thereafter, the biodiesel prototype will undergo procedures of refinement and determination of physical properties, adapted to the available time framework of the project. The students learn from their mistakes and failures by designing and applying a prototype product, having the potential of continuous monitoring redesign.

In the third phase the activities involve the environmental footprint of biodiesel. It is about a set of activities that lead "biodiesel" to the boundaries of science-society relation. Investigating the environmental effects of biodiesel use through the life-cycle assessment (LCA), the stages in which the benefits in biodiesel use to CO2 emissions occur, will be recorded and assessed. During this phase, the issue of indirect land use change (iLUC) will also be studied, which raises doubts about the benefits of biofuels use. The definition of the phenomenon of iLUC and its social effects as well as the hitherto conflicting research findings from the modelling of phenomenon will give students the chance to view science not as certain knowledge but as a constant attempt of approaching the optimal solution to a problem, meeting a basic goal of science education, that of formulation of appropriate attitudes and perceptions towards science.

REFERENCES

Banks, F., & Barlex, D. (2014). *Teaching STEM in the secondary school: Helping teachers meet the challenge*. Routledge.

Barakos, L., Lujan, V., & Strang, C. (2012). *Science, Technology, Engineering, Mathematics (STEM): Catalyzing Change Amid the Confusion*. RMC Research Corporation, Center on Instruction.

Burrows, C. A., Breiner, J. M., Keiner, J., & Behm, C. (2014). Biodiesel and integrated STEM: Vertical alignment of high school biology/biochemistry and chemistry. *Journal of Chemical Education, 91*(9), 1379–1389. https://doi.org/10.1021/ed500029t

Drake, M. S., & Burns, R. C. (2004). *Meeting standards through integrated curriculum*. ASCD.

Hmelo-Silver, E. C. (2004). Problem-based learning: What and how do students learn? *Educational psychology review, 16*(3), 235–266. https://doi.org/10.1023/B:EDPR.0000034022.16470.f3

Laboy-Rush, D. (2011). *Integrated STEM education through project-based learning*. http://www.rondout.k12.ny.us/common/pages/DisplayFile.aspx.

Maryland State STEM Standards of Practice (Draft). Accepted by the Maryland State Board of Education April 2012.

McCormick, R. (2006). Technology and knowledge: Contributions from learning theories. In J. R. Dakers (Ed.), *Defining technological literacy: Towards an epistemological framework* (pp. 31–47). Palgrave Macmillan.

Moore, J. T., & Smith, K. A. (2014). Advancing the State of the Art of STEM Integration. *Journal of STEM Education: Innovations and Research, 15*(1), 5–10.

Norström, P. (2013). Engineers' non-scientific models in technology education. *International journal of technology and design education, 23*(2), 377–390. https://doi.org/10.1007/s10798-011-9184-2

Ontario Ministry of Education (2002). *The Ontario Curriculum, Grades 11 and 12: Interdisciplinary Studies*. Queen's Printer for Ontario.

Sanders, M. (2009). STEM, STEM education, STEM mania. *Technology Teacher, 68*(4), 20–26.

Smith, A. K., Douglas, T. C., & Cox, M. F. (2009). Supportive teaching and learning strategies in STEM education. *New Directions for Teaching and Learning, 117*, 19–32. https://doi.org/10.1002/tl.341

Tseng, K. H., Chang, C. C., Lou, S. J., & Chen, W. P. (2013). Attitudes towards science, technology, engineering and mathematics (STEM) in a project-based learning (PjBL) environment. *International Journal of Technology and Design Education, 23*(1), 87–102. https://doi.org/10.1007/s10798-011-9160-x

Wiggins, P. G., & McTighe, J. (2005). *Understanding by design.* Ascd.

Chapter 3

STEMART = STEM + ART: An Erasmus+ STEAM Project for K-12 Education

Milan Hausner[1], Paraskevi Karypidou[2], Maria Gratiela Popescu[3], Serkan Altinöz[4], Laura Bozzo[5], Olya Mihova[6], & Petra Mikše[7]

[1]*Zakladni skola Praha 3, Lupacova 1/1200, Czech Republic*

[2]*1st Primary School of Diavata, Thessaloniki, Greece*

[3]*Scoala Gimnaziala nr 7, Buzau, Romania*

[4]*Topalli Ortaokulu, Antalya, Turkey*

[5]*Fedac, Manresa, Spain*

[6]*General Secondary School "Nikolay Katranov", Svishtov, Bulgaria*

[7]*Osnovna Sola Jozeta Krajca, Rakek, Slovenia*

Abstract: *The STEMART project focuses on European educational priorities, mainly the support of STEM education. Seven countries (Czech Republic, Greece, Romania, Turkey, Spain, Bulgaria, Slovenia) will organize seven scientific weeks and competitions on chosen topics, to which an international approach will be shown. There are scientific explorations, technical inventions, biological experiments, programming and also construction activities. A permanent sharing of experiences, creation of a scientific ledger and various outputs gives students new horizons and perspectives. As can be clearly seen, a precise evaluation of outputs will create an exceptionally rich scientific and technological experience which can have a strong effect on the future job orientation of students. All products will be published under Creative Commons License. This approach will also involve national bodies, institutions and political representatives in the dissemination of this project. The project will be organized under the auspices of the Czech commission of UNESCO and the representative of European parliament.*

Keywords: *STEM, STEAM, education, Erasmus+, K-12*

1. Introduction

The priorities which have been chosen are essential to the future of Europe because they create active roles for prospective citizens and also support and strengthen their own motivation, challenges and involvement. In a time when STEM has become one of the most important values in European education the aim should be to develop other priorities, which are creativity, motivation, cooperation and team work on an international scale. All of these priorities are part of the partner school framework, but this project will give them new multicultural and social impacts. Because this project will be the focus of all school partners, it will also become the flagship of international education in all municipalities in partner countries.

The STEAM (STEM with ART) project is an Erasmus+ project for K-12 institutions with duration of two years, starting on September 2016. The project is coordinating by the Czech school with six more partners from six countries (Czech Republic, Greece, Romania, Turkey, Spain, Bulgaria, Slovenia).

STEM (Science, Technology, Engineering and Mathematics; Delaney, 2014; Feldman, 2015), is an acronym that refers to the academic disciplines of science, technology, engineering and mathematics. STEM education combined with Arts education (STEAM) should provide us with the education system that offers us the best chance for regaining the innovation leadership essential to the new economy (inGenious 2016; Jolly 2016; Science, Technology, Engineering and Mathematics, 2016; STEMALLIANCE, 2016).

The challenges of our generation will demand creative solutions, so innovation in art and sciences needs to be encouraged. For example all European schools are currently going through some important changes. The old system is not related to the work field so current curriculum need to be revisited and rearranged. It is essential to update our educational system because employment opportunities are different. The schools have to change as soon as possible. Teachers should find the students' learning styles in order to adapt teaching methods to the pupils' needs. Teachers are expected to personalize lesson scenarios or work differently with students while considering the learning style of the students (by groups). All of these priorities could create an exceptionally rich project with a large impact on the future existence of partner schools.

Another priority which is fully included in the STEAM project is the fact that all partner countries significantly contributed to STEM knowledge for mankind and to present these endeavors to the younger generation. One example comes from history (from Turkish aqueducts to the viral remedies of modern Czech scientists). In STEAM typical and unique national products will be used to show how STEM will contribute to them. Such examples could be the unique Mercur construction set or Spanish programming applications.

Skills in Science, Technology, Engineering and Maths (STEM) are becoming an increasingly important part of basic literacy in today's knowledge economy ("Science, technology, engineering, and mathematics," 2016). According to the EUN Schoolnet webpage (Science, Technology, Engineering and Mathematics, 2016), to keep itself growing, Europe will need one million additional researchers, technologists and mathematicians in applied science by 2020 (European 2020 Initiative). The creativity of the international teams, sharing of own and mutual values, each school's own work in the field of STEM, mutual offline and online meetings together with development of digital content all of these skills in one project creates a fantastic challenge not only for students but also for teachers. When discussed with municipalities in all countries, full support and a will to cooperate were given. The project will involve all parent organizations and parents of students in schools, as without them it wouldn´t be possible to fulfill all the goals which are the following. This approach will support another aspect of modern Europe the family bonds which are being so dramatically destroyed in these modern times. Another aspect is international understanding and support for European mobilities. Europe has recently been facing the migration crisis with a lot of prejudices, xenophobia and racism. Cooperation with various nations in one group, creating work together and their sharing of it could be a real asset for such a goal. It will not overcome all barriers, but it could help the way children approach the issue of migration.

In this paper, Section 2 introduces the goals and methodology of the project. Section 3 presents the main implementation principles and planned outcomes of the project. Section 4 outlines the expected impact and Section 5 briefly mentions some of the planned dissemination actions and expected exploitation of the project results.

2. Goals and Methodology

The methodology of the STEMART project mainly focuses on supporting challenges and activities targeted to student groups and team collaborative learning, in various environments of the participating schools and in each location. This will bring together education with polytechnic real tutoring. There is a list of activities in partner schools which will be provided both online and offline. Other important features of the project are regularity and complexity. Each partner will prepare their own program of activities with focus on the international level. This organizational model will also be a part for the evaluation board of partners, who will check procedures and protocol of the project.

The opening staff meeting (Prague) is an essential part of project. It will provide a necessary start to all other activities and outcomes as benchmarks, quality and methodology plans are to be discussed. Recognition of the personal motivation of all school project managers is the second must. This meeting will open not only professional cooperation, but also personal bonds. At this meeting the situation of the school regarding social status, location and possibilities will be discussed and evaluated from various aspects.

The second staff meeting (Buzau, Romania) will have the goal of evaluating works done so far and create the possibility to improve results achieved. It is also a good place for retrospective evaluation and an opportunity to discuss a precise plan for dissemination.

Since all staff in schools will have to be involved, it is essential to give them appropriate challenges and motivation. The extent of outcomes can't be attained without the vast support of teachers and school community.

The project has clearly set and quantified outputs, so it will be easier to evaluate overall quality metrics. Quality assurance will also be introduced into the system by a board of evaluators. This quality assurance will especially evaluate the impact of the outcomes of students for all schools.

Daily activities in all partner schools will provide motivation and appropriate challenges for involvement in international cooperation. That is why the workshop topics cover all STEM parts (Science, Technology, Engineering and Math by virtue of ART) and the variety of options will ensure that the majority of the students will be able to find a number of activities where they can be involved. All student outcomes will be presented to their respective local school communities and the best of them will be shown in the international project conferences.

3. Implementation and Outcomes

Because each school is fully responsible for the complex development of student personality, there also has to be support of the sense of art and culture as essential parts of life. Some of the STEM activities could also be seen through DIALOGUE OF ART WITH STEM.

Students and teachers in all partner schools will choose one topic of interest from STEM as the priority for the school, region and country. The list of such priorities is listed below. The school together with the local community will prepare for 5 days of school workshop activities to show a complex view of the topic chosen. It will involve students, teachers, parents, municipal politicians and also of course local institutions working in the area. This conference will be organized in partner schools and the school will create all possible support for other partner schools in this field. They will organize webinars, online discussion, creative writing, an art contest and all imaginable activities to create a challenge for the students in this field. Such workshops will be organized in all partner schools.

The partner school community will develop support for topics offered in a way that at least 10 assets (clip, art work, multimedia presentation) will be included in every topic. They will organize an indoor school contest with the participation of local institutions and also small school mini-conferences to choose 4 participants for the international conference in a partner school. It is expected that about 70% of students in the age range of 10-12, 12-14 will somehow be involved.

As the project sets 100 mobilities as the limit which it supports, every partner has to choose at least 4 conferences for their presentation internationally. All schools will endeavor to find other sources of finance in order to participate in more than three conferences.

3.1. Learning, Teaching and Training Activities

These activities will also be introduced in a flipped classroom. Workshops will be held in all partner schools and led by students as well as by various experts from the country of the organizer. 7-week workshops will have a mini-conference, terrain work, excursions, socializing, challenge games as well as various discussions with experts on chosen topics. The value of such learning is based on the diversity of environments, various ages of presenters, and most valuably on 'hands-on' activities. Visits to real lab environments and enterprises facilitate students and teachers to attain a new in depth view into real scientific, artificial and business life. These activities are the core of this scientific and art project as it facilitates the creation of the international working framework and gives all students full access to European shared youth knowledge.

Czech Republic: "*3D virtual science models*": A set of workshops focused on modeling through construction sets. Visits to science laboratories and museums. Streamlined pupil's conference about 3D models. Presentation of models of partner countries. Cultural events with historical and political presentation. Visits to the regional town hall and socializing activities.

Romania: "*House full of experiments*": Set of scientific experiments which will be provided by students from the host and from visitors. Cultural and political socializing together with a streamlined conference to the other participants. Visits to local monuments.

Spain: "*Programming Code*": Presentation of pupils Android and Win applications. Special workshop for Google (Android) programming. Presentation of applications of students. Contest for the most efficient and attractive Android application. Cultural and social meetings. Streamlined conference and official publication of the applications.

Turkey: "*Historical inventions*": Presentations by pupils about historical inventions in their home country, visit to the museums and workshops about the historical inventions. Presentation of national inventors and his/her endeavors for recent and future time. Streamlined conference. Presentation of Scientific Ledger and Database of inventors and scientists. It will be organized with the help of Mediterranean University together with cultural and social events in Antalya.

Slovenia: "*Art under microscope*": Presentation of microscope images made by pupils in all partner schools, Art exhibition of microscopic structures. Workshops for students from other countries. Presentation of parts of their cultural heritage. Organization event supported by local media, local representatives and school ministry. Streamlined workshop to the other partners.

Bulgaria: "*Architecture around us*": Presentation of typical architectonic styles from partner countries, visit to the national monuments and comparative workshop about architecture. Presentation of various architectonic styles and streamlined conference. Presentations of 3D models derived from various construction sets. Models could not only be typical buildings, but also technical endeavors such as bridges, viaducts, roads etc.

Greece: "*Lego Mindstorm robotics*": Exhibition and workshop based on Lego programming. Workshop based on Mindstorm Lego programming for students from both schools. Presentation of applications by the visitors. Streamlined conference. The cultural and social part of the visit is connected to the history and local landscape. Presentation of results to the political bodies.

3.2. Expected Project Outcomes

There are various outcomes planned. School workshops are not mentioned here as they are listed below. General products are as follows:

1. LOGO of project the first international contest will be focused on development of a logo for the project.

2. STEM QUEST multilateral questionnaire with prizes in both cultural and STEM topics which will show partner schools and also represent the national approach in this field.

3. STEM CALENDAR the International calendar presenting STEM achievements for all partner countries (at least 12 events per every partner).

4. STEM VIP PERSONS the international multimedia database for at least 15 persons whose achievements are important both for STEM and country.

5. 7 STEMbyART TRAVELING EXHIBITION in the last three months there will be this three-day exhibition both for students and school community. Because of complicated logistics a part of the exhibition will be done by video-stream from the partners responsible for the topic shown. 3D objects will be presented in the partner school which is responsible for the theme.

6. INTERNATIONAL STEM JOURNAL web-based journal about new achievements in STEM which will be edited by students. Every partner is responsible for creating 70 references during

the project time. All of these references will be tagged and cross referenced. It will consist of scientific, technological, engineering and mathematical achievements at national levels.

7. STEM YOUTUBE CHANNEL Partner schools will edit its own YouTube channel with the STEM topic where all videos will be published and i-framed for http://stem.lupacovka.cz

8. 8 MINI-CONFERENCE PROCEEDINGS WITH STUDENT, GUESTS PRESENTATIONS. These products will be available at least one month after the school's mini-conference. Every proceeding will have at least 70 articles. Proceedings will be shown only online and no printouts will be published.

9. SCHOOL STEM WEB LEDGER Work on the project could be followed by a web ledger which is already available at http://stem.lupacovka.cz.

10. CULTURE AND NATURE POSTCARDS every partner will prepare introductory spots about their country, region, school and participants (One presentation per country video on YouTube channel.)

11. CREATIVE COMMONS LIST OF COPYRIGHTS All participants (students and teachers) involved in the project will get a CERTIFICATE OF COPYRIGHTS to show their personal copyrights to the mental works. Such an approach will support activities which try to overcome copyright abuse by personal experience.

3.3. Accessibility of Project Outcomes

Because all products will be created in the framework of project, all products will be accessible under Creative Commons License Attribution 4.0 International (CC BY NC 4.0) which allows: adaptation, remix, but not commercial use of the products (Creative Commons 2016).

As the target group is meant to be mainly K-12, there is no official instrument to validate the results of the conference. The team of evaluators will consist of experts on a national level and these experts will evaluate results of students and all proceedings by their statements. Such statements will be the essential part of project. There will be the international board of experts who will declare their statements to the online proceedings of all conferences.

A special award for the best participants in all workshops will be given together with a list of copyrights (Creative Commons) to all authors of student products.

Another evaluation will be created by mutual questionnaires and polls among students who will participate in the online and offline workshops. Evaluation will be the vital part of all 7 proceedings from the workshops. Preformatted questionnaires will be published.

Because the project is based on STEM, the project will be certified by a listing in the EUN STEM project on the web: //http://www.eun.org/focus-areas/stem. This portal will be the essential reference and dissemination tool on the European level.

4. Impact

The most expected impact is for the target group to increase their motivation for STEM subjects, as these subjects were very often reduced in comparison to the teaching of languages. The real use of language in another subject (CLIL) is the other possible impact. A project of this scale requires full participation of all school staff because of the diverse topics included in the project. It could support school team work, as it is well-known that teachers are very often individualists and that sharing of content is not always their first priority. The sharing of experience and materials is another impact for the future. The questionnaire at the start of project will be repeated and statistic verification of the improvement of student attitudes to STEM and to polytechnic subjects is expected.

Students will learn that STEM creates the future and language connects people. This is a simple definition, but for the motto of the project, it is essential.

4.1. Impact Assessment

As was mentioned above, all results have quantitative and qualitative benchmarks which will be discussed in the final report. The number of publicized articles and discussions will be one of the most important impacts. Questionnaires for evaluation of attitudes to STEM and project satisfaction will evaluate the impact on participants and on the target groups. A group of experts will publish their own statement at the national level (politician, a teacher from another school, scientist, and journalist). Awards for best presentations and children's products will be given and published, so all the European community will have access to all 7 workshops results as the main part of the dissemination activities. A visitor counter with detailed Google analytics will be implemented, so at the end of the project a complete analysis of the results will be available.

This project shows other ways of cooperation when compared with Comenius. It is built on real scientific and educational activities with clearly described outcomes and products. In addition, recognition of not only cultural and historical aspects, but also of STEM issues, is included in many identification benchmarks in all countries and contributes to the to the advancement of the international understanding and cooperation among the project partners. According to earlier research, no other project with such activities has so far been organized, and all outcomes will define unique educational materials, which can be shared via European http://lre.eun.org as the main European portal for learning objects.

At the national level, the project will be introduced into the national LO object contest called DOMINO http://domino.nidv.cz and http://dum.rvp.cz. All materials will be published in all national portals for learning objects. The project will have strong media support at the regional and national level and it will also be introduced to EUN bodies and to the social media.

One of the desired impacts of the project, even before its launch, is a new concept of school club which will be open during summer 2016 and is called *"STEM school club"*. The special school club for kids aged 6-11 will be established and organized by ZŠ Lupáčova and these premises will also be used for some project activities in the next two years. A view of this club can be seen on http://kreativita.lupacovka.cz.

All results will be published and disseminated at both national and European levels as a project of such versatile content is completely unique and follows all European priorities. In all partner countries the official dissemination links will be adopted and carefully checked for partner share in this field.

4.2. Sustainability

The project aspires to create a new school community based not only on professional cooperation, but also on personal bonds created during two years of mutual work. A group like this ensures a possible new common creativity. Results published in the school portal will give everybody new challenges. A large number of outcomes and all activities are expected, in which such content was created can easily be repeated and new threads to the event could be added. The creation of a Gallery of scientists and Science Ledger offers the chance to create new items and local follow-ups of these products. The Polytechnic education STEM, as was also mentioned, has become a part of a new School club concept and the output of students could be another motivation. 3D Merkur virtual models can also create challenges for mechanical engineering in the class (Merkur Toys s.r.o, 2016).

5. Dissemination and Use of Results

The first target group for which the results will be shared with is the teachers' community in the educational portals of EUN. Among them the most important ones are:

a) http://lre.eun.org, where results will be shared under Creative Commons license.

b) http://www.eun.org/focus-areas/stem will be the essential European community where in all results will be introduced and shared.

c) In accordance with Mr. Poche, representative of European parliament, this project will also be introduced to the Youth panel in: http://www.europarl.europa.eu/european-youth-event/en/take-part!.html.

The main section of results will be streamlined and published via You Tube. They will also be offered for publication on the SCIENTIX EUN portal (SCIENTIX 2016), both for final and intermediate evaluation and assessment purposes. All countries will also individually discuss with NA publication in recommended sources and conferences.

Because of the vast content and complexity intermediate project results are to be presented in December 2018 at EMINENT/SCIENTIX conference.

Some of the planned national dissemination activities are listed below:

Czech Republic - In school: peer program in other classes, Erasmus Days + corner, PTA meetings, presentation to the representative from the Town Hall.

Publication in local and national media. The UNESCO ASP School will present the project on the UNESCO associated school network as well as on the member of Asia Europe Classroom Network, for partners in this circle.

1) http://rvp.cz - National educational portal where examples of best practices in education can be published

2) http://praha3.cz - Local portal of Prague 3 municipality

Greece - Erasmus Days meeting with briefings at school, diffusion within the school, update school website, update teacher community afternoon meeting with parents. Municipality of Delta complex presentation to local bodies. Local TV channel. Participation in the:

3) http://www.dimosdelta.gr - Local web portal of Delta municipality

4) http://www.sch.gr - The Greek School Network portal

5) http://dipe-v-thess.thess.sch.gr - West Thessaloniki Elementary School Admin. portal

Turkey - Erasmus Days in school and community, social events with local politicians and parents. Presentation of project to local administration and to local media.

    6) http://aksu07.meb.gov.tr/ School webpage

    7) http://www.antalyaaksu.gov.tr/ Local governmental webpage for region

    8) http://topalliortaokulu.meb.k12.tr/ Local school portal

Romania - The school's magazine: A spy in the school; meetings with other teachers from other schools in our town/ county; local TV and Local newspapers, presentation to the local administration, PTA meetings

    9) http://www.scoala7bz.webs.com school site

    10) http://www.primariabuzau.ro/ official administrative portal of Buzau, Romania

Spain - Erasmus School Days, PTA meetings, School Board presentation, local feasts and children's presentation. Meetings with other teachers during professional CPD.

    11) http://www.regio7.cat/ this is the web of the county's newspaper

    12) http://www.fedac.cat/novafedac/ web of 25 Christian Fedac Schools

    13) http://canaltaronja.tv/bages/ local TV mainly focused for regional news

Slovenia - Erasmus School Days+ corner, Local municipality gathering, PTA association meetings and also local newspaper

    14) http://www.cerknica.si/ local administration official webpage

    15) http://osrakek.si/ - school website

Bulgaria - Erasmus School Days+ corner, Local municipality gathering, PTA meetings, presentation to the Teacher association in Bulgaria

    16) http://svishtov-news.com/ local info server

    17) http://www.beta-iatefl.org/ Bulgarian teacher association

REFERENCES

Creative Commons. Retrieved October 30, 2016, from https://creativecommons.org/

Delaney, M. (2014, April 2). Schools Shift from STEM to STEAM. *EdTech Magazine.* http://www.edtechmagazine.com/k12/article/2014/04/schools-shift-stem-steam

Feldman, A. (2015, June 16). STEAM Rising: Why we need to put the arts into STEM education. *Slate Magazine.* http://www.slate.com/articles/technology/future_tense/2015/06/steam_vs_stem_why_we_need_to_put_the_arts_into_stem_education.html

inGenious: the European Coordinating Body in Science, Technology, Engineering and Maths education. Retrieved October 25, 2016, from http://www.ingenious-science.eu

Jolly, A. (2014, January 9). 5 Hot Topics in STEM Education. *Middleweb.* http://www.middleweb.com/17350/five-hot-topics-stem-issues

Merkur Toys s.r.o. Retrieved October 26, 2016, from www.merkurtoys.cz

Science, Technology, Engineering and Mathematics. *European Schoolnet.* Retrieved October 20, 2016, from http://www.eun.org/focus-areas/stem

Science, technology, engineering, and mathematics. (2016, October 6). In *Wikipedia.* https://en.wikipedia.org/wiki/Science,_technology,_engineering,_and_mathemati-cs

SCIENTIX. Retrieved October 20, 2016, from www.scientix.eu

STEMALLIANCE. Retrieved October 12, 2016, from http://www.stemalliance.eu/

Chapter 4

Searching for Black Holes. Photometry in Our Classrooms.

Ioannis Chiotelis[1] & Maria Theodoropoulou[2]

[1]*Experimental High School of University of Patras, University of Patras Campus, 26504, Rio Patras*

[2]*Department of Primary Education, University of Patras, 26500, Rio, Patras*

Abstract: *Following the main axis of STEM, this module integrates the use of inquiry-based learning methodology and the integration of ICT tools in schools practices. Is devoted to the implementation of a research-based experiment where students can be involved in the identification of stellar mass black hole candidates and the procedure to "measure" their mass limits. We encourage students to operate the Salsa-J software, perform some photometry measurements and try to spot a Black Hole candidate through images captured by telescopes, available on line, or ordered to be taken through a remote telescope. At the end of this module students should know how to identify black hole candidates and how to determine the mass limits of a compact object in a binary star system. Thus, Science, Physics and Astronomy are strongly supported by ICT technology and Mathematics is contributing the data processing and conclusions outcomes.*

Keywords: *Photometry, black holes, Salsa-J, binary star systems, full width half maximum*

1. Introduction

1.1. Binary Star Systems

Since 1970 when UHURU satellite was launched by NASA, the study of binary star systems was rapidly developed (Belczynski et al., 2002; Blanchet & Schafer,

1993; Postnov & Yungelson, 2006; Thorsett et al., 1993). Especially, X-ray radiation emitting binary systems are of high importance, while in this case one of the components is a compact object probably a black hole or a neutron star, and the other component a 'normal' star (usually a main sequence star or red giant) (Aharonian et al., 2005; Gallo, Fender, & Pooley, 2003). The star usually orbits around the common center of mass gradually losing its mass towards the black hole or neutron star. Thus, the discovery of a binary X-ray star system can reveal an out of sight black hole.

While the compact object is pulling matter from its companion due to the intense gravitational field a disc is forming around the compact object called accretion disc. Depending on the position of the companion star and the compact object, with its accretion disc, different amounts of light are coming towards the observer (Fabian et al., 1989). Actually, we cannot observe the individual components, but only a dot which brightness changes in time. It is from the study of these changes in the form of a light curve that we can conclude to some of the binary's system characteristics. The image in Figure 1 shows the infrared light curve of the black hole candidate (Shahbaz et al., 1994). Depending on the position of the companion star and the compact object, with its accretion disc, we see different amounts of light coming towards the observer. Thus, we can assume whether the compact object is a black hole or a neutron star.

1.2. Photometry

Photometry gained significant importance recently as it is used to discover exoplanets by measuring fluctuations in the intensity of a star's light over time (Brown et. al., 2001).

Generally, photometry is used to generate light curves revealing the variability of light output over time of objects such as variable stars and supernovae. Photometry is the measurement of the intensity or brightness of an astronomical object (e.g., star or galaxy). For example, a star looks like a point of light when you look at it just with your eyes but the Earth's atmosphere smears it out into something that looks like a round blob when you use a telescope to look at it. In order to measure the total light coming from the star, we must add up all of the light from the smeared out star.

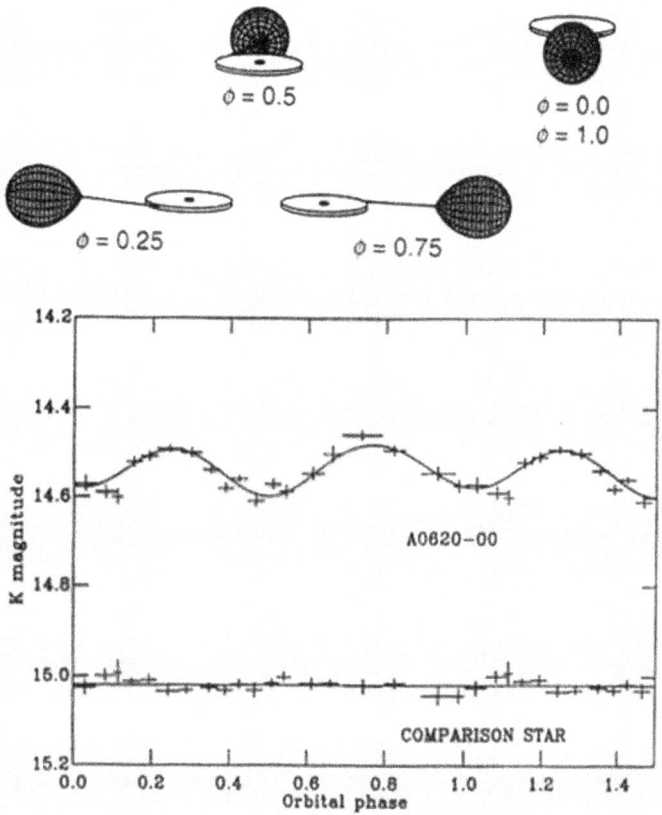

Figure 1: Infrared light curve of a strong black hole candidate.
Source: Shahbaz et al, 1994.

1.2.1. Aperture Radius

One of the most important definitions in photometry is the aperture radius. Aperture radius defines the radius of the circle that is used to count the pixel values in the image. The radius of the circle is very important if the radius is too small, it will not count all the light coming from the star and if it is too big, it may count too much background sky or other stars in the image.

Therefore, we may not get accurate measurements. In our software (Salsa-J) the radius of aperture is automatically set as the Full Width Half Maximum (FWHM) of the stars in the image. The FWHM is used to describe the width of an object in the image. The FWHM is an average measure that counts the telescope's optics, the image recording CCD and the atmosphere through which the light passes. In order to estimate the FWHM is to cross sect a star image (Figure 2).

The base of the slice is at the value for the background sky, while the top of the peak represents the counts measured in the brightest pixel along the slice. The FWHM is the number of pixels across the peak at a point halfway up from the base.

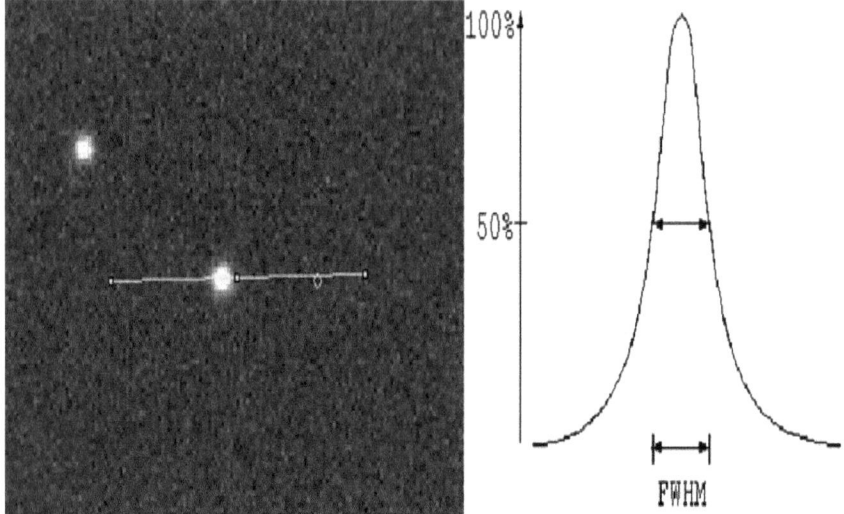

Figure 2: Estimation of Full Width Half Maximum (FWHM)

2. Experiment

2.1. Calibration

In order to calibrate our photometry measurements we must follow the next steps. In Salsa-J, we choose Analyse>Photometry Settings. Then we select force Star Radius and choose the value 6. Next, we go to Analyse>Photometry and another empty window will then appear entitled 'Photometry' (Figure 3). Using the mouse, we click on a star in our image. The intensity of the object is calculated by adding up all the pixel values within the radius of the aperture. We can clearly see that the chosen radius is too small. Let's increase it a bit to 8. Click on the star again and a new measurement will appear. Still the radius is too small. Increase the value to 10 and continue until you reach a radius of 40.

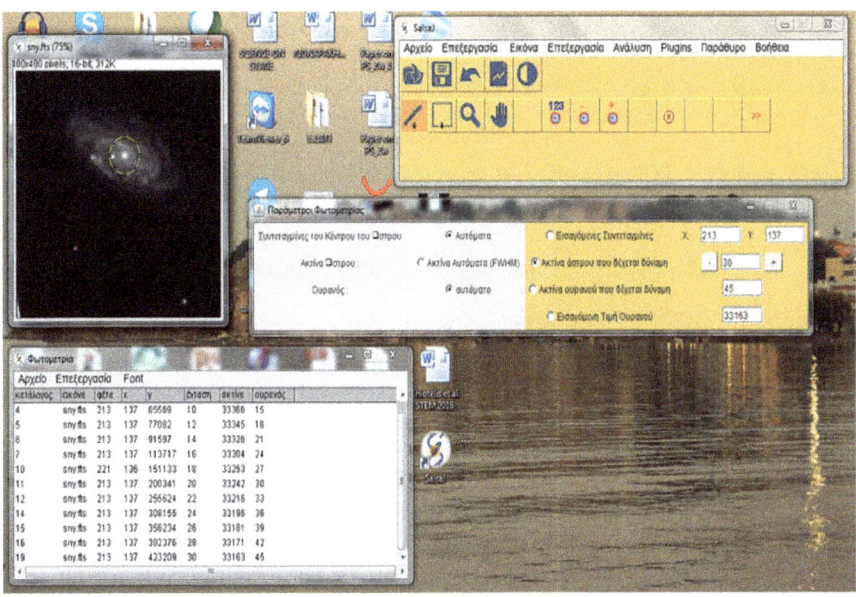

Figure 3: Calibrating the photometry measurements

Then in Excel we create two columns one for Radius and one for Intensity adding the radius and intensity values from Salsa-J. Then we plot a graph of this data (Figure 4).

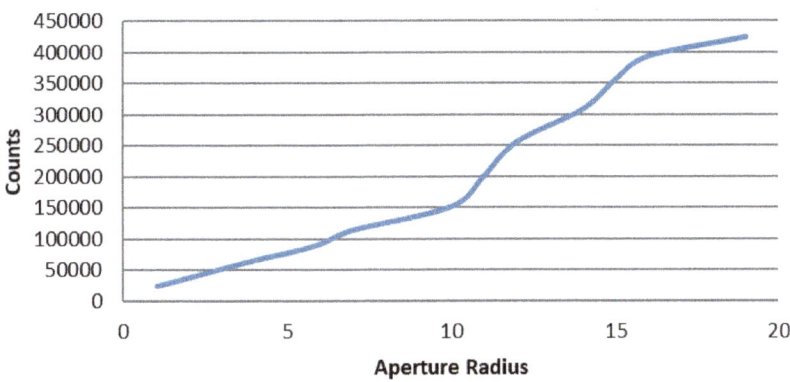

Figure 4: Photometry calibration. Counts for Aperture Radius

We can see the rapid rise of intensity as the radius of the aperture increases. This is because more of the star is included in the increasing radii of the apertures. The graph begins to flatten out when we have the entire star within the aperture, but keeps rising gradually as more and more of the background sky is included. From this graph, we can see that the best radius to use is around 20. Once we have chosen the best aperture radius, this can be set for our photometry analysis. It is advisable to carry out this exercise every time we come to work with a new set of images. These steps are the main steps of photometry calibration and photometry measurements.

2.2. Finding A Black Hole Candidate

2.2.1. The Study System-Object

The object we will study is the black hole candidate XTE J1118+480. It was discovered in March 2000 by the Rossi X-ray Timing Explorer satellite. It is approximately 6000 light-years away in the constellation of Ursa Major (Figure 5).

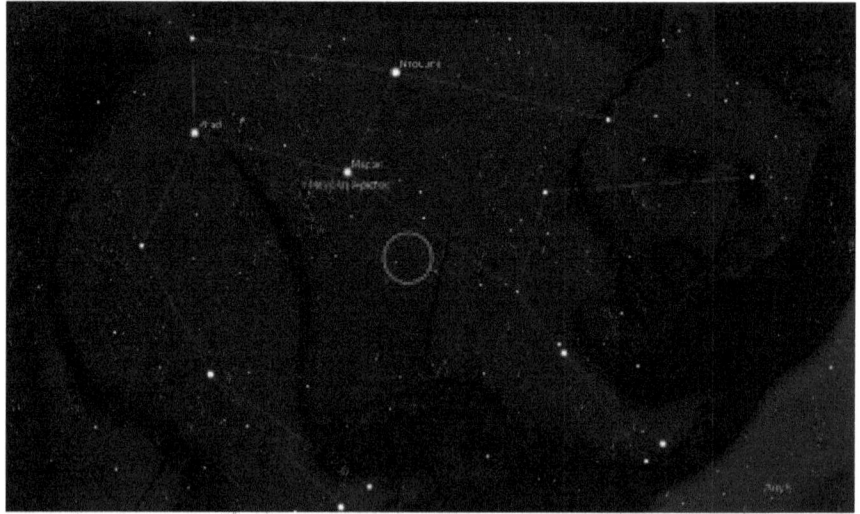

Figure 5: Location of the black hole candidate XTE J1118+480 (Credit: Stellarium)

The system is composed of a compact object and a less than 2 solar masses star. The optical component of this system is a star (KV UMa), while the estimated mass of the black hole candidate is around 7 solar masses. This is precisely what we want to confirm with this exercise. The data we analyzed were obtained by Faulkes Telescope (https://archive.lco.global/ on13/05/2009) and we followed the photometry analysis procedure using Salsa-J.

2.2.2. The Experimental Procedure

As we can see on Figure 6 several stars are surrounding the object we wish to study. We will select some of these stars as reference stars in our study. The procedure is to make photometry measurements of all these stars and the black hole candidate in as many as possible images (Doran et al., 2012). We are looking for differences in the brightness (intensity) of the comparison stars. These variations probably reveal that these stars are orbiting around the accretion disc of the black hole candidate. In any case we must be careful while these variations can often be attributed to weather conditions during the observation.

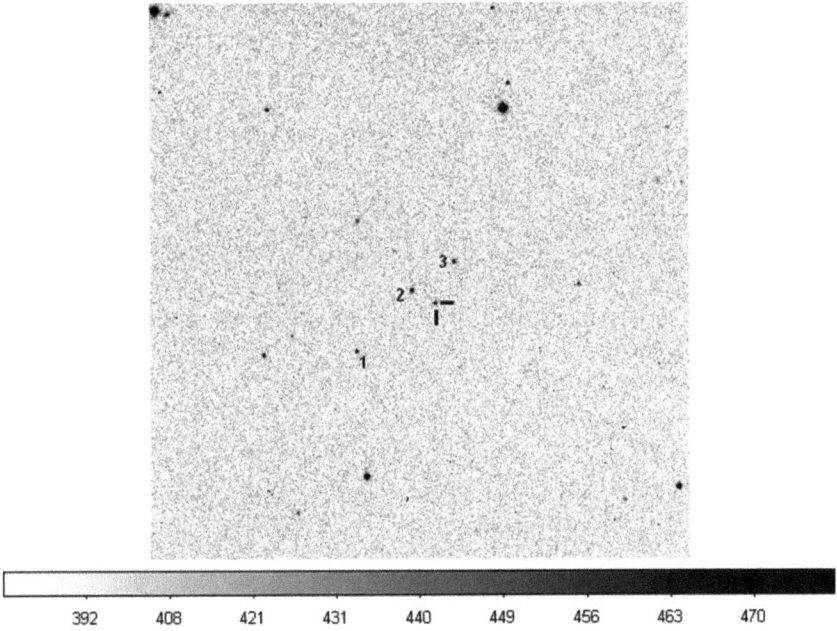

Figure 6: The map of stars locating the object and the comparison stars. XTE J1118+480 is denoted by the two black lines the comparison stars are shown as 1, 2, 3.
Source: Faulkes Telescope North

Ensuring that the variations are due to orbital reasons we can then plot the intensity against time. Thus we should trace the variability and use this to estimate the mass of the non-visible compact object. According to the procedure described in paragraph 1.2.1 (Figure 2) we are determining the best aperture radius before proceeding with the photometry. In practice, a good choice for the radius of an aperture is about 1.5 or 2.0 times the FWHM (Figure 7).

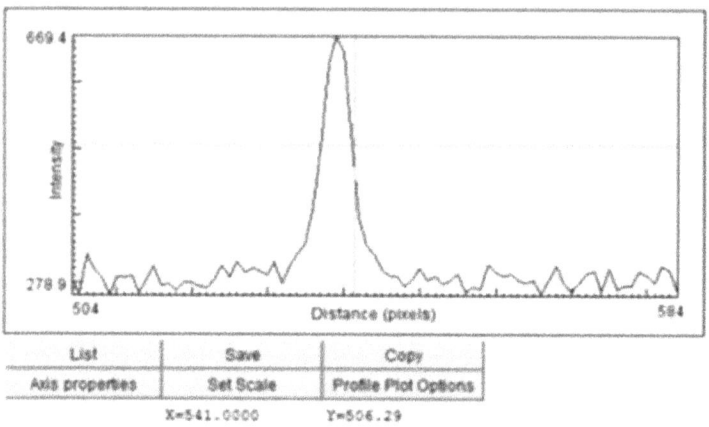

Figure 7: Evaluating the best aperture with Salsa J

Then we must measure the intensity for the 3 comparison stars and for the black hole candidate in as many as possible different images. During our school project, we distribute the images amongst a group of students, ensuring that each group uses the same aperture radius and the same comparison stars. Furthermore, we adjust the brightness and contrast in all images in order to be able to see all three objects. If we can't see all of them we decide not to use the particular image.

2.2.3. Data Processing

Since we are working with relative magnitudes, we don't have to worry about absolute magnitudes and standard stars. We are not looking for the absolute value of the magnitude of the object but the variations to its intensity relative to other stars in the same image. We also used the Modified Julian Date (MJD) for each image. We found this information in the header of FTS images. In Salsa J we selected the "Show Info" under the Image menu and in the header we found the value for MJD. This is the value to be used on the x-axis of our graph. Thus, we plotted the following graph (Figure 8):

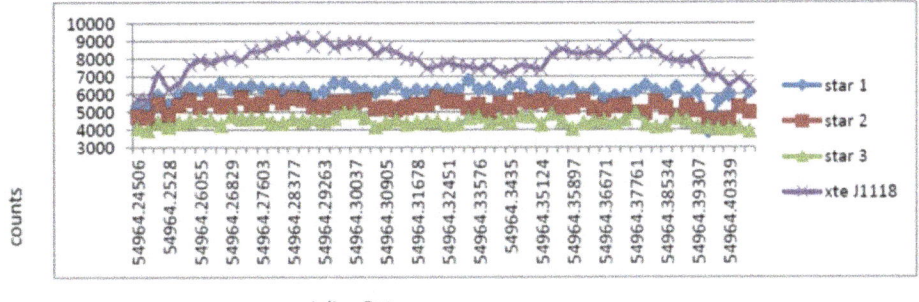

Figure 8: Plot of counts vs. time

From this graph we can clearly see that our target (xte J1118) varies far more than the comparison stars (stars 1, 2, 3). We already know that the orbit of the visible star and compact object around each other is periodic. Finding the orbital period is complicated and time demanding (Doran et. al., 2008). But we can make a rough estimate from the values plotted on Figure 8. The formula that transforms the Julian dates in Phase is the following:

$$Phase = \frac{MJD - T_0}{Period} \quad (1)$$

Where the MJD is given in the header of each image and T_0 is the MJD of the first image. Scientists already know the period of this object P= 4.08 hrs = 0.17 days. (http://adsabs.harvard.edu/abs/2001ApJ...556...42W). The results are being shown in Figure 9.

3. Results

Using the following formula we tried to determine the mass limit of the compact object (Lewis et. al., 2008).

$$f(M) = \frac{M_1^3 \sin i^3}{(M_2 + M_1)^2} = \frac{P K_2^3}{2\pi G} \quad (2)$$

where M1 and M2 are the masses of the compact object and the companion star respectively, P the orbital period, G is the universal gravitational constant, i is the inclination of the orbital plane of the system with the line of sight of the observer and K2 the radial velocity of the visible star.

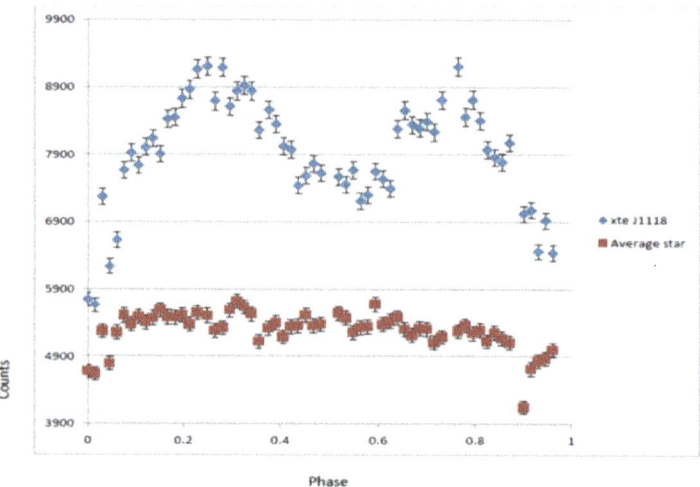

Figure 9: The orbital period of our system

According to scientists (http://adsabs.harvard.edu/abs/2001ApJ...556...42W) the radial velocity of the visible component of this system was determined to be ~700 km/s and the mass of the companion is ~6.1 Solar Masses. We tried to confirm these values from our measurements. Calculating that P = 0.17 * 24 * 60 * 60 = 14 688 s and Msolar = 1.9891 × 1030 kg we have (Roberts et. al., 2013):

$$f(M) = \frac{M_1^3 sini^3}{(M_2 + M_1)^2} = \frac{14688s * (\frac{700Km}{s})^3}{2\pi * 6{,}67384 * \frac{10^{-11}m^3}{Kg*s}}$$

$$f(M) = \frac{M_1^3 sini^3}{(M_2 + M_1)^2} = 1.2 * 10^{31} Kg \sim 6.3\, M_{solar}$$

Although we used approximate values, we can mention that our calculations reached a very good value for the mass of the stellar black hole candidate XTE J1118+480. (http://arxiv.org/pdf/astro-ph/0104032.pdf)

REFERENCES

Aharonian, F., Akhperjanian, A. G. Aye, M. K., Bazer-Bachi, A. R., Beilicke, M., Benbow, W., & Bolz, O. (2005). Discovery of very high energy gamma rays associated with an X-ray binary. *Science, 309*(5735), 746–749. https://doi.org/10.1126/science.1113764

Belczynski, K., Kalogera, V., & Bulik, T. (2002). A comprehensive study of binary compact objects as gravitational wave sources: evolutionary channels, rates, and physical properties. *The Astrophysical Journal, 572* (1), 407–431.

Blanchet, L., & Schafer, G. (1993). Gravitational wave tails and binary star systems. *Classical and Quantum Gravity, 10*(12), 2699–2721.

Brown, T. M., Charbonneau, D., Gilliland, R. L., Noyes, R. W., & Burrows, A. (2001). Hubble Space Telescope Time-Series Photometry of the Transiting Planet of HD 209458. *The Astrophysical Journal, 552*(2), 699–709. https://doi.org/10.1086/320580

Doran, R., Lobo, S. F., & Crawford, P. (2008). Interior of a Schwarzschild black hole revisited. *Foundations of Physics, 38*(2), 160–187. https://doi.org/10.1007/s10701-007-9197-6

Doran, R., Melchior, A. L., Boudier, T. Ferlet, R., Almeida, L. M., Barbosa, D., & Roberts, S. (2012). Astrophysics data mining in the classroom: exploring real data with new software tools and robotic telescopes (arXiv:1202.2764.). *arXiv*.

Fabian, A. C., Rees, M. J., Stella, L., & White, E. N. (1989). X-ray fluorescence from the inner disc in Cygnus X-1. *Monthly Notices of the Royal Astronomical Society, 238*(3), 729–736.

Gallo, E., Fender, R. P., & Pooley, G. G. (2003). A universal radio–X-ray correlation in low/hard state black hole binaries. *Monthly Notices of the Royal Astronomical Society, 344*(1), 60–72.

Lewis, F., Russell, M. D., Fender, P. R., Roche, P., & Clark, J. S. (2008). Continued monitoring of LMXBs with the Faulkes Telescopes (arXiv:0811.2336.). *arXiv*.

Postnov, K., & Yungelson, L. (2006). The evolution of compact binary star systems. *Living Reviews in Relativity, 9*(6). https://doi.org/10.12942/lrr-2014-3

Roberts, S., Roche P., & Lewis, F. (2013). Educational projects with the Faulkes Telescopes. In A. Lazoudis (Ed.), *Proceedings Discover the Cosmos Conference: e-Infrastructure for an Engaging Science Classroom* (pp. 25–32). EPINOIA S.A.

Shahbaz, T., Ringwald F. A., Bunn, J. C., Naylor, T., Charles, P. A., & Casares J. (1994). The mass of the black hole in V404 Cygni. *Monthly Notices of the Royal Astronomical Society, 271*(1), 10–14.

Thorsett, S. E., Arzoumanian, Z., McKinnon, M. M., & Taylor, J. H. (1993). The masses of two binary neutron star systems (arXiv:astro-ph/9303002.). *arXiv*. https://doi.org/10.1086/186758

Chapter 5

Developing a Design Framework for UMI Educational Scenarios

Olga Fragou[1], Achilleas D. Kameas[1,2], & Ioannis D. Zaharakis[1,3]

[1]*Computer Technology Institute and Press, Diophantus, Patras, Greece*

[2]*School of Science and Technology, Hellenic Open University, Patras, Greece*

[3]*Computer and Informatics Engineering Department, Technological Educational Institute of Western Greece, Greece*

Abstract: Ubiquitous learning (u-learning) is a new paradigm which is based on ubiquitous computing technology. The most significant role of ubiquitous computing technology in u-learning is to construct a ubiquitous learning environment which enables anyone to learn at any time anyplace. Nonetheless the characteristics of u-learning are still unclear and being debated by the research community. Designing instructional tools that actually promote u-learning experiences is a cumbersome task in the sense of taking into consideration and combining a variety of complex, technological tools and characteristics of u-learning. UMI stands for ubiquitous computing, mobile computing, Internet of Things. This study describes the characteristics and design methodology of a UMI-Sci-Ed Educational Scenario Template as a medium to organize and construct u-learning experiences based in a u-learning environment. It also presents a case study scenario, based on UMI Subject Matter Experts' interaction with the predefined and designed Educational Scenario Components.

Keywords: *u-learning, instructional design tools, multidisciplinary educational scenarios*

1. Introduction

The evolution of ubiquitous computing has been accelerated by the improvement of wireless telecommunications capabilities, open networks, continued increases in computing power, improved battery technology and the emergence of software architectures (Liyytinen & Yoo, 2002). Its emergence creates new conditions for all working as educational professionals and learning as students. However, the key factor is not the logic or technical specifications of the machines but the new ways in which meaning is created, stored delivered and accessed (Cope & Kalantzi, 2009, 173). Just using state-of-the-art machines in learning, does not mean that new learning is taking place. The new paradigm of u-learning, emerging of these technology developments has to be based in an instructional paradigm and context which makes use of the advanced characteristics of the new technology affordances, towards shaping up a robust framework of u-pedagogy. Under this scope, this paper presents important characteristics of u-learning paradigm and technologies. Section 2 describes the basic characteristics of u-learning while Section 3 includes the proposed approach in developing a design framework for u-learning educational scenarios, highlighting design characteristics. Sections 4 and 5 detail the methodology of the design framework and present as a case study, preliminary data of a scenario design in UMI-Sci-Ed project, formed with the contribution of SMEs (Subject Matter Experts) in Internet of Things (IoT) technologies, in Higher Education. The conclusions are presented in the final part of the paper.

2. U-learning

Currently, u-learning is carried out in various educational settings and investigates in different directions such as ubiquitous pedagogy, classroom centered u-learning mode, specific curriculum centered u-learning mode, faculty education for the implementation of u-learning, development standards of u-learning resources and development of u-learning instructional management system (Zhang, 2010). Just because the computing is ubiquitous, not all learning has to be machine mediated, and distanced from its natural and embodied sources: the machines need to be seen, not as ends in themselves but as documentation devices for off-screen learner activity. Even though u-learning has attracted the attention of researchers, the criteria or characteristics for the establishment of u-learning are still unclear (Hwang et. al., 2008). U-learning seems to integrate best characteristics of former pedagogical paradigms as presented in the Figure 1.

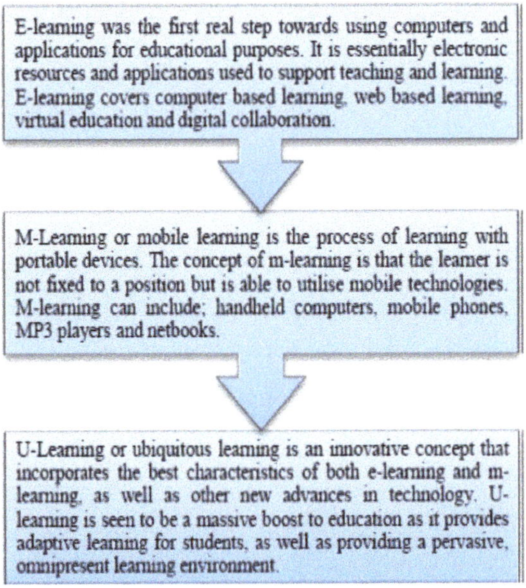

> E-learning was the first real step towards using computers and applications for educational purposes. It is essentially electronic resources and applications used to support teaching and learning. E-learning covers computer based learning, web based learning, virtual education and digital collaboration.

> M-Learning or mobile learning is the process of learning with portable devices. The concept of m-learning is that the learner is not fixed to a position but is able to utilise mobile technologies. M-learning can include; handheld computers, mobile phones, MP3 players and netbooks.

> U-Learning or ubiquitous learning is an innovative concept that incorporates the best characteristics of both e-learning and m-learning, as well as other new advances in technology. U-learning is seen to be a massive boost to education as it provides adaptive learning for students, as well as providing a pervasive, omnipresent learning environment.

Figure 1: Evolution from e-learning to u-learning

Six characteristics of m-learning have been adapted by various researchers as part of u-learning: urgency of learning need, initiative of knowledge acquisition, mobility of learning setting, interactivity of learning process, situating of instructional activity, and integration of instructional content (Koper & Specht, 2008). Though the most prominent characteristics of u-learning are permanency, accessibility, immediacy, there is one another parameter for considering the learners' mobility within the embedded computing environments concluding on more two characteristics: interactivity and situating of instructional activities (Hwang et. al., 2008). Utilizing context-aware and ubiquitous computing technologies in learning environments encourages the motive and performance of learners (Koper & Specht, 2008). Under this scope, as main characteristics of u-learning emerge: urgency of learning need, initiative of knowledge acquisition, interactivity of learning process, situation of instructional activity, context awareness, actively provide personalization services, self-regulated learning, seamless learning, adapt the subject contents, and learning community (Chiu et al., 2008). Thus, u-learning does not come without challenges, such as the need for context aware infrastructure, along with methods for development of specific tools that are based to a particular situation and the necessity to research, from a human computer interaction perspective, new paradigms of interaction with ubiquitous and contextualized media and learning experiences (Koper & Specht, 2008). Thus, designing and developing a learning environment that supports u-learning is by default a complex and demanding process which has to take into consideration not just the technological parameters but also combine and rebuild in a creative way traditional modes of learning, adhering to new learning and evolutionary situations.

3. Developing a Design Framework for U-Learning Educational

3.1. Characteristics

An educational scenario is defined as an educational setting in predefined time frame: it describes an educational arrangement designed or set up to provide a rich methodological educational unit. It aims at presenting all important and subsidiary functional aspects of learning (i.e., time, context of learning etc.). Its use in multiple educational settings and contexts aided by the use of technologies is important for orchestrating tools and processes in authentic education settings, in an effort to plan learning (Stewart, 2008). In the context of u-learning, the use of educational scenario seems to serve the following purposes:

- It complies with the interest for rich activity-based pedagogies that originate from various socio constructivist influences.

- It is tied up to the goal of creating deeper, better integrated and applicable knowledge.

- Its use, as structured schema, is required in creating effective "new pedagogies", where the teacher has to fulfil a triple role as facilitator, manager and orchestrator.

- It shapes the learning space describing the social space that provides intellectual and emotional support.

- It provides a constantly evolving structure for the design transition to describing advanced learning situations (i.e., virtual learning activities).

- It means that both the structure and content of a scenario can be used for alternate presentation mode.

On designing u-learning educational scenarios there are certain moves which explore and exploit the potentials of u-learning, shaping affordances under a different spectrum. These moves comprise important actions on designing a u-learning experience such as (a) blurring the traditional institutional, spatial and temporal boundaries of education, (b) shifting the balance of agency from the teacher to the student, (c) recognising learner differences using them as a productive resource, (d) broadening the range and mix of representational modes, (e) developing conceptualising capacities, (f) connecting one's thinking into the social mind of distributed cognition, (g) building collaborative learning cultures.

A u-learning environment relies on the constructivist learning theory; teachers just not deliver information but they also provide guidance as well as an explorative learning experience. It also relies on conversation and collaboration between students, who can learn by interacting with and even teaching other students. To capture basic components of the learning process, we initially designed a conceptual framework in the form of an educational template, comprised by several blocks (Figure 2).

Synopsis presents a brief educational scenario description. The Scenario orientation/Focus includes information describing the scenario on the basis of knowledge, skills and attitudes expected during the scenario implementation process. A brief content analysis on the basis of key terms presents basic information for scenario categorization. Shaping the temporal space of the educational scenario the block Time Distribution refers to the actual implementation time of the described educational scenario. Using the Bloom taxonomy, Expected Learning Outcomes block provides information on the basis of precise sentences describing what learners are expected to accomplish. Though u-learning scenarios are multidisciplinary, we thought it was important to place the scenarios' design on specific curriculum areas, presented by the block Placement and Course. The Actors that actually influence the learning process are the Teacher and Students on a first basis: however, in this case, social schemata of Communities of Practice are expected to function as moderators in knowledge creating and knowledge sharing. As far as the pedagogical design of the template is concerned subsidiary functional aspects of the learning process, Modes of Interaction, Delivery, Media and Products have been included as blocks. Modes of Interaction refer to the types of orchestration and organization, the desired modes of interaction between Actors. Media refer to the technological tools involved in the educational scenario implementation whereas Delivery includes on ways the students have access in the educational scenario. The block Products of the educational scenarios includes artefacts, source code and digital material produced during the learning process. The underpinning pedagogical theory is described in the Pedagogical Elements block, whereas learning prerequisites that are important to be fulfilled before implementing the educational scenario are included in this block. The Content block presents the actual content that students will see and explore during the educational scenario implementation. As Assessment is an important factor in the learning process effectiveness there has been a block introduced in the UMI-Sci-Ed Educational Scenario Template, whereas possible ways of scenario expansion, are introduced to the corresponding block. A separate block on describing ways of Reflection and Feedback has also been introduced in the UMI-Sci-Ed Educational Scenario Template.

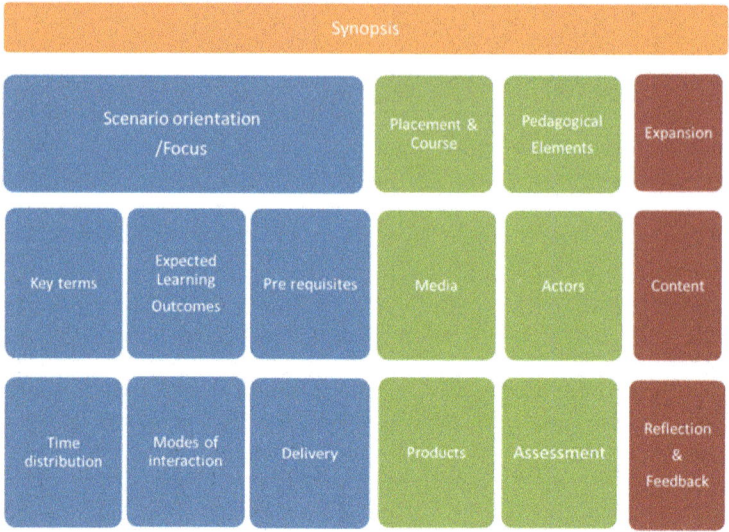

Figure 2: The UMI-Sci-Ed educational scenario template

4. Methodology

4.1. Scenario-Based Design

Scenarios follow systematic and recognizable steps. Though there are numerous techniques in scenario design, there are aspects quite similar: the designer has first to clarify the important decisions that (s) he has to take, challenging the mental maps that shape people's perceptions and collecting information from various sources. The next steps are more analytical: identifying the driving forces, the predetermined elements and the critical uncertainties. The deeper structure and system behind the scenario narration and their underlying logics are elaborated to explain them and reveal their crucial differences. Finally, the key events or turning points that would channel the future towards one scenario version are identified (Hwang et al, 2008; van Merriënboer & Pass, 2003).

The scenario-based design methodology has been adapted because (a) scenarios evoke reflection in design, (b) design situations are fluid, (c) design moves have many consequences, (d) any scenario has many views and (e) technical knowledge lags technical design. Technical professionals are experienced people performing complex tasks: they want to reflect on activities and they routinely do reflect on activities. To design the UMI-Sci-Ed Educational Scenario Template initial decisions had to be made based on content analysis so as to shape a generic content schema including important aspects of the learning process: actors, media, content, learning outcomes and assessment. On further developing the design blocks of the educational template we decided to add categories such as Key Terms, Delivery and Expansion to provide a more elaborated description in a meta-level. U-learning scenarios include activities such as (i) gathering and distribution of information, (ii) creation of collaborative learning documents, (iii) discussions and comments about the productions and (iv) project management related activities.

Constructing scenarios-of-use inescapably evokes reflection in the context of design: the scenario emphasizes and explores goals that the user may adopt and pursue. When designs incorporate rapidly evolving technologies, requirements change even more rapidly. The more successful, the more widely adopted and the more impactful a design is, the less possible it will be to determine its correct design requirements. To manage an ambiguous and dynamic situation, the design has to be concrete but also flexible. It seems that effective reflection must be tightly coupled to action (van Merriënboer & Pass, 2003): the analysis needs not be complete and consistent, it needs only to guide a restructuring of the current situation that can produce new design actions or new insights.

Powerful pedagogical designs that aim at the development of general problem skills, deeper conceptual understanding and more applicable knowledge include characteristics such as (a) the use of complex, realistic and challenging problems that elicit in learners active and constructive processes of knowledge and skill acquisition; (b) the inclusion of small group, collaborative work and ample opportunities for interaction, communication and co-operation; and (c) the encouragement of learners to set their own goals and provision of guidance for students in taking more responsibility for their own learning activities and processes (van Merriënboer & Pass, 2003).

4.2. What About STEM Education?

As the education profession develops programs to address the evolving STEM teaching and learning needs, a central factor that must be understood is that STEM content and STEM education are not the same (Stewart, 2008). Further, expertise in one STEM discipline does not automatically translate to expertise in another discipline even if the process of scientific inquiry is well understood by the learner (Froyd & Ohland, 2005). The multidisciplinary nature of STEM education presupposes conceptually generic pedagogical schemata, however open to elaboration so as to be applicable in a variety of learning situations covering an array of subject domains. Jobs go unfilled in STEM settings because there is a gap between what graduates in STEM can do and the skills STEM employers are seeking (Stewart, 2008). The problem is not about STEM graduates lacking of the technical skills needed for the job but actually the non-technical, such as ability to network, time management, the so-called "soft skills". In a research sample of 1,065 employers surveyed by the Deloitte Access Economics commissioned by the Office of the Chief Scientist, Australia, these emerged as important work skills required for STEM graduates to achieve success in their workplace (Table 1).

The importance placed on the skills and attributes varied by industry sector: for example 86% of employers in the Information, Media and Telecommunications sector rated programming important or very important-much higher than other sectors. This sector also rated design thinking as high priority.

Table 1: Respondents' rating of each of 13 different skills and attributes

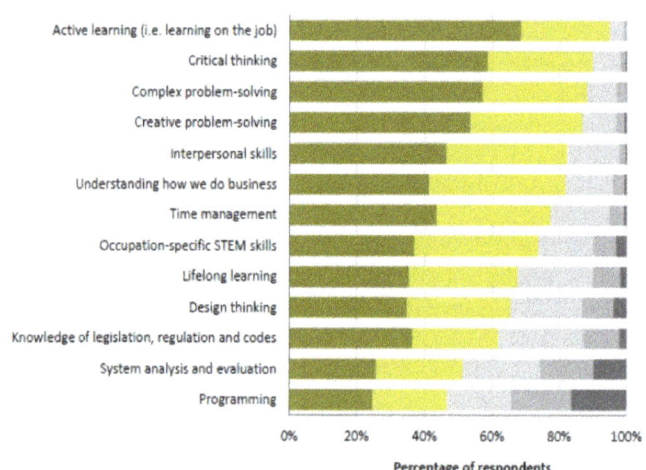

Source: Data adapted from Deloitte Access Economics Report, Australia's STEM Workforce: a survey of employers, 2014.

5. Towards Developing UMI-Sci-Ed Educational Scenarios Smes Case Study I

In order to start using the UMI-Sci-Ed Educational Scenario Template, the design of educational scenarios for 14+ youngsters using UMI technologies for STEM education has been initiated. The following section presents a brief description of the scenario designed by SME in UMI targeting at educating students in cryptography principals and concepts. The rationale of designing the scenario covered 3 basic phases as presented as follows.

Step I – Analyse the problem

Safety issues in using wireless technology and smart devices are very important in the sense that users need to be serviced by the use of UMI technologies in their everyday lives, however they also have to learn to protect themselves from risks emerging from malicious use and possible intruders. Under this scope, being able to process data in order to i) make decisions regarding safety issues and in a sense privacy issues, and ii) critically examine parameters of data entry so as to select appropriate technique for data encryption, are the basic steps in this educational scenario. The educational scenario's basic steps are presented as follows:

- Introduction of basic steps and principals of cryptography
- Selection of the application and the data that are going to be encoded
- Selection of appropriate algorithm, e.g. symmetric key algorithm
- Performing of encryption process according to selected algorithm
- Performing of reverse operation for decryption
- Outlook and knowledge feedback

Step II – Identify strengths and weaknesses

People wishing to engage in a secure exchange of information will swap public keys and use some method to ensure the existence of identical private keys. Creating an educational scenario based on concepts and processes of cryptography has proven a quite cumbersome task in the sense of pinpointing and elaborating material suitable for 14+ youngsters. However, the interest of young users in a topic like that (cryptography) and the challenge of making a subject like that feasible for learning with UMI technologies seemed to be the greatest challenge of all. The experts must explain how the security and consequently cryptography relates to privacy and how it facilitates the control of information flows (i.e., who gets to know what when?) and helps to ensure the correctness of data.Students have to be prepared for obstacles such as the proper exchange of private keys and complete transactions in a secure manner. They also have to face problems such as the compromise of a private key. By using this scenario students are expected to:

- Learn fundamental primitives of cryptography and its relation to privacy
- Familiarized with the encryption/decryption processes
- Summarize the basic issues of data encryption and decryption

- Summarize security issues of ubiquitous computing

- Identify the need of cryptography in ubiquitous computing applications as an integral part of any privacy solution

Step III – Analysis of the external environment

The time slot for applying the specific scenario is 4-5 hours in workshop settings, involving also a grouping mode (4-5 students). As hardware tools, UDOO NEO (http://www.udoo.org/udoo-neo/) can be used with the appropriate peripherals for data entry and I/O external devices. UDOO NEO is an open hardware, low-cost platform equipped with a powerful 1GHz ARM® Cortex-A9 and an Arduino UNO-compatible platform that clocks at 200 MHz, based on a Cortex-M4 I/O real-time co-processor, all wrapped into an i.MX 6SoloX processor by NXP. This is actually a tool for fast prototyping, which provides a boost to DIY (Do It Yourself) world, and a new vision to the educational framework: the idea of training up a new generation of engineers, designers and software developers skilled in digital technology: physical computing, multi-media arts, interactive arts and Internet of Things.

6. Conclusions

Design based methodology has been used to as to start developing an UMI-Sci-Ed Ubiquitous Science Learning Environment (USLE). Projects in STEM learning encompass the many foundations of STEM learning, including intellectual, behavioural and social factors, as well as emerging contexts and tools for STEM learning. Research on learning environments investigates new, high-impact learning opportunities in STEM, including in classroom, non-classroom and virtual settings. The UMI-Sci-Ed Educational Scenario Template has been presented in its initial design stage as well as preliminary findings covering the design of a ubiquitous computing educational scenario based on UDOO NEO hardware kit. The educational scenario template is currently evaluated and updated according to SME's feedback. The next step is to start developing and orchestrating samples of activities based on the actual educational scenario design so as to start expanding basic concepts and processes in a learning ecology using UMI technologies.

Acknowledgements

This project has received funding from the European Union's Horizon 2020 research and innovation programme under grant agreement No 710583.

REFERENCES

Chiu, P. S., Kuo, Y. H., Huang, Y. M., & Chen, T. S. (2008). A Meaningful Learning based u-Learning Evaluation Model. In B. P. Woolf, E. Aïmeur, R. Nkambou, & S. Lajoie (Eds.), *Proceedings 8th IEEE International Conference on Advanced Learning Technologies* (pp. 77–81). https://doi.org/10.1109/ICALT.2008.100

Cope, B., & Kalantzis, M. (2009). Multiliteracies: new literacies, new learning. Pedagogies, 4(3), 164–195. https://doi.org/10.1080/15544800903076044

Froyd, E. J., & Ohland, M. W. (2005). Integrated engineering curricula. *Journal of Engineering Education, 94*(1), 147–164.

Hwang, G.-J., Tsai C.-C., &Yang, S. J. H. (2008). Criteria, Strategies and Research Issues of Context-Aware Ubiquitous Learning. *Educational Technology & Society, 11*(2), 81–91.

Koper, R., & Specht, M. (2008). Ten-Competence: Life-Long Competence Development and Learning. In M-A. Cicilia (Ed.), *Competencies in Organizational e-learning: concepts and tools* (pp. 234–252). IGI-Global.

Liyytinen, K., & Yoo, Y. (2002). Issues and Challenges in Ubiquitous Computing. *Communications of the ACM, 45*(12), 62–65.

Stewart, M. T. (2008). Crafting interactive case studies for tertiary training. *Ako Aotearoa–National Centre for Tertiary Teaching Excellence.* Retrieved March 23, 2016, from http://akoaotearoa.ac.nz/communities/toolsdelivering-scenario-based-e-learning-both-locally-and-across-internet

van Merriënboer, J. J. G., & Paas, F. (2003). Powerful learning and the many faces of instructional design: Toward a framework for the design of powerful learning environments. In E. De Corte, L. Verschaffel, N. Entwistle, & J. G. van Merrienboer (Eds.), *Powerful Learning Environments: Unraveling Basic Components and Dimensions* (pp. 3–20). Pergamon.

Zhang, A. H. (2010, 5–6 July). Study of ubiquitous learning environment based on Ubiquitous computing. In R. Klamma, R. Lau, S.-C. Chen, Q. Li, I. Ahmad, & J. Zhao (Eds.), *3rd IEEE International Conference on Ubi-media Computing* (pp. 136–138). IEEE. https://doi.org/10.1109/UMEDIA.2010.5544482

Chapter 6

New Perspectives for Geometry Teaching: Mechanical Linkages Technology

Kalliopi Siopi[1] & Eugenia Koleza[2]

[1]*Model High School of Evangeliki School of Smyrna, Athens, kalsiopi@gmail.com*

[2]*University of Patras, Department of Primary Education, Patra, ekoleza@upatras.gr*

Abstract: *In this paper we try a brief presentation of mechanical linkages, and especially of the drawing machines. Our focus is on the pantograph, which incorporates mathematical properties and relationships in structure in such a way to allow the implementation one geometrical transformation, such as, symmetry, reflection, translation and homothety. In order to investigate subjects' concepts/ theorems-in-action developed by investigating the structure of the pantograph, and especially the identification of the math concepts and laws incorporated in the machine, we selected a pantograph's model and taught homothety to high school students for four hours (early 2016), in the framework of an attempt to incorporate artifacts with the characteristics geometrical machine's in the instruction of Euclidean geometry.*

Keywords: *linkages, pantograph, geometry, teaching, homothety*

1. Introduction

Learning with the use of various technological tools attracts a lot of research interest and holds great promise; the use of the tools in the current learning environments allows the students to have increased chances of achieving mathematical concepts, exploring and experimenting with mathematical ideas and expressing such ideas and concepts via a variety of representations (Taimina, 2008). At the same time, integrating engineering-based problem solving within the learning objectives of the students is also very appealing in the progressive societies, making the teaching of Science, Technology, Engineering and Mathematics (STEM) a very important component of their education.

Technology is often referred to as the tool and the application of math and science. Within the classroom, computer technology is often used to facilitate lesson planning, class activities or even to create resources. Studying the history of technology, during the mechanical age (between 1450 and 1840) a lot of new technologies were developed. These technologies embodied the knowledge of their time, and were the first attempt of math and science modeling. Such technologies include the slide rule (an analog computer used for multiplying and dividing) and the Pascaline (a very popular mechanical computer) by Blaise Pascal that could be used even nowadays. Nevertheless, the use of mathematical machines as technological tools to promote learning seems to be under-researched, with limited existing knowledge regarding the experience of the students. A sound exception is the MMLab (Laboratory of Mathematical Machines, www.mmlab.unimore.it) where researchers have investigated from an epistemological and pedagogical aspect the use of mechanical machines - concerning geometry and arithmetic as a way to generate mathematical ideas or concepts in the classroom (Bussi et al., 2010; Martignone, 2011; Mariotti et al., 1997). This Research Group has investigated the use of simple mathematical machines in different contexts and grade levels. These machines (for example pantographs) are linkages that allow the implementation of geometrical transformations, such as symmetry, reflection, translation, and homothety. Unlike some artifacts from ICT (Information Computer Technology), they need to be handled appropriately, require motor abilities, might resist the motion and need time to be explored. The role of mathematical machines notably those that are part of the historical phenomenology of geometry, as a rule, compasses, mechanisms and geometrical machines (for example the pantographs), and their didactic exploitation in activities are the subjects of research studies and their use is discussed in teaching activities in the classroom (Mariotti, 2002; Maschietto, 2005; Maschietto & Bussi, 2011).

The use of simple machines was under investigation also in the SIMALE project (the Simple Machines Learning Environment) aimed to support the mechanical reasoning and understanding of middle and high school students. The project was created to support "reflection, collaboration, and presentation of concepts from multiple perspectives and contexts" and resulted in the development of a learning environment (McKenna & Agogino, 2004).

In this paper, we focus on geometrical mechanical linkages in order to explore their nature and their role in geometry teaching. More specifically, we focus on the pantograph as an example of didactic exploration of a simple machine model in classroom activities about similarity.

2. Machines and Mechanisms

One of the most fundamental examples of machines is the linkages. A mechanical linkage is a series of rigid links connected with joints to form a closed chain, or a series of closed chains by having a link or links fixed, and by means of which straight or nearly straight lines or other point paths may be traced. Each link has two or more joints, and the joints have various degrees of freedom to allow relative movement. Mechanical linkages are usually designed to take an input and produce a different output, altering motion, velocity, acceleration, and applying mechanical advantage. If two or more links are movable with respect to a fixed link then a linkage is called a mechanism.

Linkage design is often divided into three categories of tasks called motion generation, function generation and point-path generation, respectively (McCarthy 2000). The point-path generation category is a classical problem in linkage design where the primary concern is the generation of straight-line paths, some examples shown in Figure 1. One of the simplest examples of linkage is the four-link mechanism or four- bar linkage mechanism. A variety of useful mechanisms can be formed from a four-link mechanism (as in Figure 2), with slight variations, such as changing the character of the pairs, proportions of links, etc.

Mechanical linkages found great applicability during the Industrial Revolution. The advances in the fields of mathematics, engineering and manufacturing processes set the ground and the need to create new mechanisms; what nowadays appears to be a simple and obvious mechanism, requires the contribution of some of the greatest minds of the time such as Leonhard Euler, James Watt (as in Figure 1a) and Pafnuty Chebyshev (as in Figure 1b).

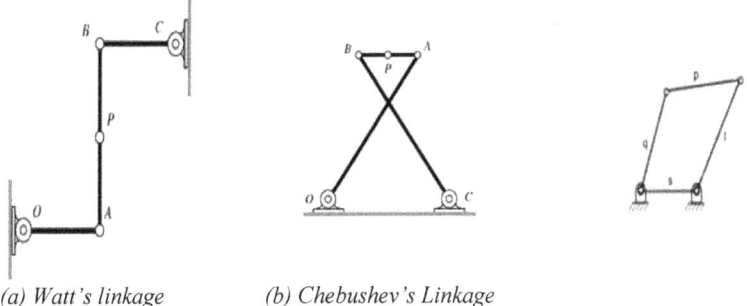

(a) Watt's linkage *(b) Chebushev's Linkage*

Figure 1: Designs of Linkages to generating Straight-line Motion

Figure 2: Four-Bar Linkage Mechanism

2.1. Mechanisms Designed by Geometric Structure

The main aspects of geometry today emerged from four strands of early human activity that seem to have occurred in most cultures: art/pattern, building structures, navigation/stargazing, motion in machines (Taimina, 2008). These strands developed more or less independently into varying studies and practices that eventually from the 19th century on were woven into what now call geometry (Henderson and Taimina 2005).

Geometry and motion came close together for ancient engineers and mechanics has been developed along with mathematics. In ancient Greece, Archimedes, Heron and other geometers used mechanisms and gears to solve geometrical problems; trisecting an angle, duplicating a cube and squaring a circle are problems that could easily be solved with the use of motion. One of the first mechanical solutions was offered by Menaechmus; he constructed a mechanical device that would trace two appropriate conic curves. Plato criticized this mechanistic approach and called instead for a purely theoretical solution with geometrical tools, as the rule and the compasses. Descartes (1596-1650), in his 'Geometry' (1637), discussed that it is obscure for him that the ancients wanted to discuss geometric construction only using circle and line and provided a response to the question 'how to make geometric constructions that they cannot be constructed with ruler and compass'. He changed the roles of ruler and compasses from exclusive geometrical tool for geometry to one of mechanical tools which were used for representing mathematics and his view enhanced exploration of mechanics and influenced the technological and mechanical sciences and the cognition (Isoda & Matsuzaki, 1999).

2.2. Special Linkages: The Drawing Machines

Before the computer era, mathematicians tried to design physical mechanics and some kinds of mechanics like figure 3, 4 and 5 still have roles as geometrical tools or drawing machines whose the mechanical structure is represented by a geometric structure, such as for example, the devices of Figures 4 and 5.

Figure 3: Ellipsografos of Descartes (1637)

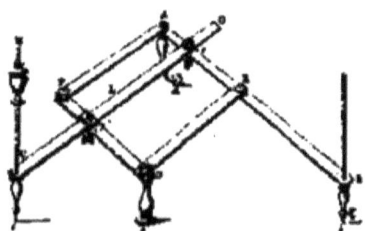

Figure 4: Pantograph of M. Bion (1709)

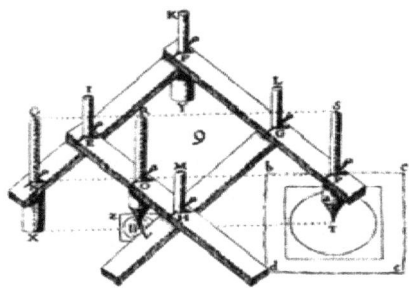

Figure 5: Pantograph of Scheiner (1603)

Drawing machine is any device/apparatus/mechanism/ instrument that draws or assists in the act of drawing. Since the early 1400s, scientists and inventors have created devices to assist in drawing. And this is not limited to "drawing realistically from life." Prisms attached to microscopes, gears and linkages joining forces for complex geometrical drawings, and intricate, specialized drafting tools are also drawing machines.

Referring to a virtual mechanism, (the mechanism model), Liu et al. (2012) distinguish between two parts of the mechanism model. (1) The information model which is used to describe the data structure of mechanism and provide information for mathematical model analysis, and (2) The mathematical model used to describe the mathematical expression of mechanism. Mechanism is composed by two parts: links and kinematic pairs, so the information model of mechanism should contain these two basic data structure. The links of the mechanism are connected through the kinematic pairs, so the motion of the links is limited in order to make the mechanism move along the route which is designed by the operator. So, the mathematical model of mechanism should be used to describe the constraint relationships among the links which belong to the mechanism. For example, the movement of a machine can be described by the geometric representation as a locus that being 'written' (as a kinematic geometry of movement). Even in mechanics, one can select between the many mathematical representations; in some cases, real situations can be more accurately described by geometrical representation and not algebraic ones. By choosing appropriate link lengths and coupler point locations, useful curves can be found, which are formalized by applying geometry to the analysis and synthesis of machines.

2.3. Special Drawing Machines: The Pantographs

The pantograph is a special drawing machine. This is an articulated system of rods which are connected with links forming simple geometrical shapes, such as similar or equal triangles, parallelograms and rhombus. The pantograph is a geometrical machine, a tool that forces a point to follow a trajectory or to be transformed according to a given law embedded in the structure of the machine (Bussi & Maschietto, 2008) and that incorporates mathematical properties and relationships in structure in such a way to allow the implementation one geometrical transformation, such as, symmetry, reflection, translation and homothetic e.g. Some kinds of pantographs' shown in Figure 6. The links allow the rods to rotate, which makes the tool not static in form and to the magnitude. The hinge/joint allows the transformation of the form of the machine and the change of the angles, without making changes in lengths and the relationships between them. The drawing-point writes locus that the proof may became within Euclidean geometry as well as that of analytic geometry. This is a good example to in order the students know the difference between of the two systems in mathematics (Bussi et al., 2010).

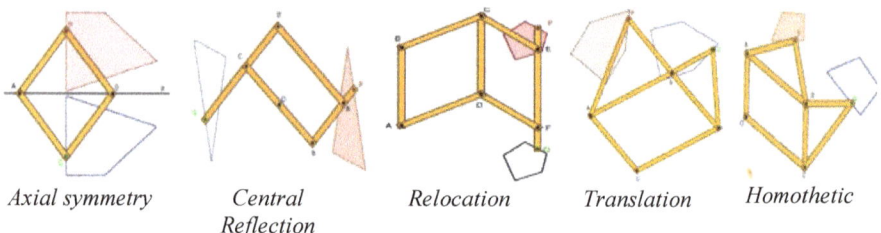

Axial symmetry　　　*Central Reflection*　　　*Relocation*　　　*Translation*　　　*Homothetic*

Figure 6: Pantographs' Kinds (http://www.mmlab.unimore.it)
Source: ***http://www.mmlab.unimore.it***

2.4. A Special Pantograph for Homothety

The idea of a pantograph is based on geometrical proportions that were known since ancient times. This drawing instrument is used for copying drawings and maps to a different scale, or in a different role for guiding a cutting tool in a manufacturing process. Gabriel Koenigs (1858–1931) expressed well the complex links between pure and applied mathematics, referring to a pantograph for homothety:

> *The theory of linkages is supposed to start in 1864. Surely linkages were used also earlier: a dedicated and precise scholar might track down them in the most ancient times. One might discover in this way that each age has in hand, so to say, yet without awareness, the discoveries of future ages: the history of things often anticipates the history of ideas. When in 1631 Scheiner published for the first time the description of his pantograph, he certainly did not know the general concepts contained as germs in his small instrument; we claim that he could not know them, as they are linked to the theory of geometric transformations, that is a theory typical of our century and gives a unitary stamp to all the made advances* (Bussi et al., 2010, p. 26).

In the case of the pantograph, like figures 4, 5, and 7 similarity is kept by the geometric structure which is based on the parallelogram and the line through the fulcrum, the force point and the influence point. In the shapes of the Figures 7 above the quadrilateral AYBC is a rhombus and AX=BZ and both are equal to the side of the rhombus. The result is that no matter how the linkage is moved X, Y and Z will always be in a straight line. If the point X is fixed (fixed-point), the point Y traces around an object (traced-point), then the point Z (drawing-point) make a copy enlargement of the object with scale factor equal ZX/YX. For example, the following schematics show that, in Figure7 the pantograph has been recommended to produce an enlargement with scale factor equal 2 because Z is always twice as far from X as Y is, and the Figure 8 the scale factor of the enlargement is 4, because X, Y, Z stay in a straight line and YZ=3XY, so making ZX=4YX (Bolt, 1991).

> *If students know the structure of the pantograph, they must change the ratio of similarity as parameters. If not, they could explore the pantograph by changing parts and discover the conditions which keep the similarity. Cognitive structures must be strictly difference before and after knowing the structure. Before knowing the structure they enjoyed changing parts but after knowing the structure there were parameters which should be changed... We should also know that same mechanics can be manipulated according to one's intuition which depends upon user's knowledge of mathematics. It should not forget that mechanics helps to develop mathematics. For example, Descartes' intuition about curves could not easily sheared with us without using his instrument* (Isoda & Matsuzaki, 1999, p. 115).

The manner of handling and use of the mechanism depends on the intuitions of the person who handles and knowledge of mathematics which he holds. Perhaps the best way to see how the pantograph works is to see its bars as part of a trellis of rhombuses, the kind you can buy in garden centers for training plants on, as representing the Figure 9 (Bolt, 1991). With this image in mind it is possible to identify other ways of making linkages to produce an enlargement. In each of these, if X is the fixed point and an object is traced out with Y, then the image traced by Z will be an enlargement from X with a linear scale factor of 3. If, however, Z traces out the object and the pencil is put at y, the image will be smaller and an enlargement from X with linear scale factor j.

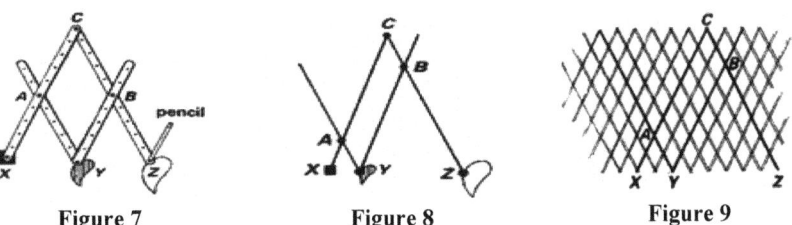

Figure 7   Figure 8   Figure 9

But the story does not end there, for we can take either Y or Z as the fixed point and in each case be left with two choices for where to put the pencil. In all the above, it should be taken into consideration the fact that the mechanistic mathematical model is able to explain its own kinematics, as a theoretically causal relation.

However, in other sciences, like social or partially engineering, the mathematical model can be discussed only by the function that fits the data in the best possible way, but the model fails to explain the actual meaning of the function as a theoretically causal relation. This is the reason behind the fact that mechanics are preferred for teaching mathematical modeling (Isoda & Matsuzaki, 1999).

3. Working with a Pantograph in A Secondary School

In order to investigate the teaching potential within the STEM framework, we selected to incorporate a pantograph's model for homothety in inquiry-based activities that refer to similarity and its' properties, to high school students for four hours (early 2016). Twenty-six students (16 years old), who had no prior experience with any artifact except for compasses and rulers, were asked to work with the pantograph that was a version of Scheiner's pantograph (as Figures 10 and 11 shown). Building blocks of our the pantograph's model were two wooden equal rods 30cm long (OD=AE=AC=BD) which were held together by the links/pivots (A, L, D, K) in the means of them forming a parallelogram (ALDK). The rods had notches allowing reassembly of the linkage while maintaining its properties provided that the links are placed in such a way that the ratios of the distances of the parts created by them are equal in each rod. The linkage had two pencils in the links A and B. The pantograph's linkage was mounted on a wooden platform (60cmx60cm).

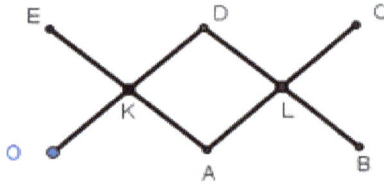

Figure 10: A photo of linkage   *Figure 11:* Schematic representation of linkage

Why this is a research STEM?

Within the STEM framework, the use of such a mechanism attempts to bridge the gap between two basic principles of the fields of mechanics and mathematics, such as cause and effect (from the mechanics side) and scale, proportion and quantity (from the mathematics side), with core ideas from the theory of geometry like parallelism, proportion, homothety and any related with them concepts (similarity ratio, enlargement). The way that this is moving forward is by engaging the students in scientific and mechanic practices, like drawing and researching, data analysing, use of mathematics (theory of geometry), discussions and planning of solutions, building arguments based on data and evaluating the information at hand.

Phase 1: Exploration of the linkage: Taking into consideration that the students had no former experience with any other tool except of the ruler and compasses, the investigation of its function was set with open, leading, questions: The articulated system that you have in your disposal, has the end of a rod adhering to the table (i.e., the end O like in the figure 11). How has it been constructed? How does it work? What can it do and why?

Do background research: The students 'read' the tool and investigate characteristics of its structure (the constrains, the length of robs, the geometrical figures, properties and relationships representing a configuration of rods, etc). They recognize that "the rods form the geometric shape of rhombus, because they have equal sides as they are the halves of equal parts", or "isosceles triangles are formed with equal sides being the halves of the rods as well as the rods, if we imagine that every second edge they are adhering to the side", or "this ratio is 1:1 so, this [OB] will be twice from this [OA]" (see Figure 11).

Identify variables: The students, experiment with the transformations of the tool's shape, taking advantage of the ability of the links, investigate the movement of the elements of the structure and they identify mobile and stable characteristics (Figure 12). Some examples: "every moment for every two [the opposite rods] are parallel, the distance of the mental parallels changes when it [the linkage] opens, but also the angles change", "when we open it the distance changes, when it is zero then this [the linkage] will be a rod with double the length of just one [the rod], when the top edges close, then this will be a rod with double the length of one rod", "when we open it, the distance between them [of the rods] increases, but they are always equal and the inside is formed, which is a rhombus or a square, but it can also be a rectangular if we change the screws".

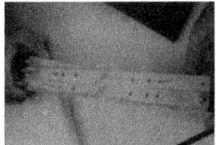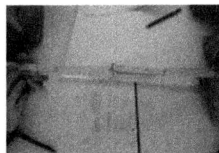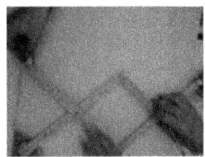

Figure 12: Transformations of the tool's shape

Formulate their hypothesis: The students experimenting with the movements of the structural elements of the tool begin to feel that some of the movements are bound and are related with the movement of some other elements. This feeling leads them to formulate some hypothesis like: *"These three [O, A, B] are in the same line in any occasion"* (hypothesis 1). *"Hypothetically, when the link A moves into a straight line, then B moves in a parallel line"* (hypothesis 2). *"When the link A describes a τμήμα, then B will form one double the size"* (hypothesis 3).

Design experiment, establish procedure: The patterns produced by the pencils are inaccurate and do not provide enough certainty for the validity of the students' claims. This leads them to devise of plans to verify their truth: *"let's place the ruler below the big triangle, wherever we extend the articulated system the two mobile edges will form a line that passes through the fixed edge* [the one fixed at the table]*"*, *"above all we need to show that* [whatever is formed] *is a straight line"*, *"through the parallel of the rods we have to show that the equal angles are moved"*, *"for the various places of B we have to show that there is something fixed, the fixed that we have is that B's distance from A is fixed as it is the distance from A to O"*. They mathematicise each situation and verify the truth of their claims using arguments based on the theory of geometry (Figure 13). As a result, they discover the relationship between the properties that are due to the structure of the tool and to the mathematical law that is implemented using the tool in any occasion (in this example shapes in magnification).

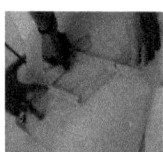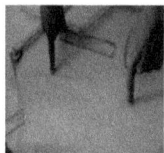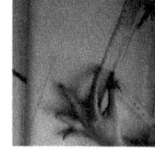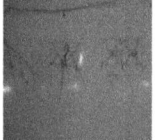

Figure 13: The plans for validation

Phase 2: Construction a special linkage's. In this phase, it was asked to construct a tool that forms shapes in a specific scale: Use the structural elements of the tool and form another one that produces shapes with the ratio 1:4. (Figure 14)

Checking and redesigning: Although they discuss about the half of the half, the connection between one of the pair of rods is done in the position 1/3 of the rod and they proceed in the verification of the connection's function as a tool that produces 4-times larger shapes, and in particular, a triangle with 4-times larger sides. They are taken aback when they realise that *"ah, this shape **cannot move with the same flexibility as the previous one"**.* They search for ways to treat the situation: *"Since we do not have any elements for this tool, maybe we could make elements for the new one, measure it, find its' dimensions and find the ratios?"*. They realise that *"this is not in a rectangular. We want to have a rectangular so this should be inserted somewhere here we must have a rectangular as well as this special* [ratio 1:4]*"*.

 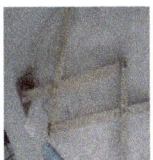 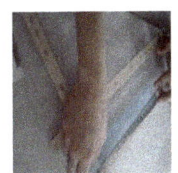 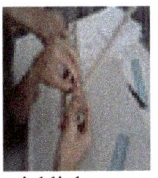 

Figure 14: The redesigning of special linkage

4. Some Remarks

Planning devices and tools can facilitate teaching of geometry as an authentic framework that, from one hand, incorporates in its' structure theoretical elements of geometry and provides information for the mathematical law that describe the way it functions (before/after teaching the relevant geometric notion, while from the other side, it improves practices (learning processes) of science, mechanics and mathematics. Simple mathematical machines in classroom activities, by reason of their characteristics and the way in which these shape and limit the possibilities of interaction with mathematical objects, regulate the production and learning of mathematical concepts. The physical and functional characteristics of the tool and the type of mathematics it incorporates, allows students to reach geometric conclusions based on the mathematics of the artifact, and may support cognitive processes focusing either on the structure of the machine, or to the embodied math concepts that emerge from the machine's movement. Technology enables us to explore situations as Machine Engineering and Art and within mathematics itself more easily. The example of the linkage for homothety the designs that produced from the point-graph on works on paper-pencil environment are incorrect. It is not expected to emerge in a simulation within Dynamic Geometry Environment, where all the objects are drawn correct. Information Computer Technologies (ITC) are not surrogates for concrete objects; rather they have their own place in mathematics education, because of the features that are partly different from the ones of physical artifacts. In the field of Education, mathematics content itself will not change radically but the value of contents could be changed depending on what and by what means should be taught.

REFERENCES

Bolt, B. (1991). *Mathematics Meets Technology*. Cambridge University Press.

Bussi, M. G. B., & Maria G. (2010). Historical Artefacts, Semiotic Mediation and Teaching Proof. In H. Gila, J. Hans Niels, & P. Helmut (Eds.), *Explanation and Proof in Mathematics: Philosophical and Educational Perspectives* (pp. 151–167). Springer. https://doi.org/10.1007/978-1-4419-0576-5_11

Bussi, M. G. B., Maria G., & Maschietto M. (2008). Machines as Tools in Teacher Education. In S. Llinares & O. Chapman (Eds.), *Tools and Processes in Mathematics Teacher Education, The International Handbook of Mathematics Teacher Education* (2nd ed.,Vol. 2, pp 183–208). Brill-Sense

Bussi, M. G. B., Taimina, D., & Isoda, M. (2010). Concrete Models and Dynamic Instruments as Early Technology Tools in Classrooms at the Dawn of ICMI: From Felix Klein to Present Applications in Mathematics Classrooms in Different Parts of the *World*. *ZDM International Journal on Mathematics Education, 42*(1), 19–31. https://doi.org/10.1007/s11858-009-0220-6

Henderson, D. W., & Taimina, D. (2005). *Experiencing Geometry: Euclidean and Non- Euclidean with History*. Pearson Prentice Hall.

Isoda, M., & Matsuzaki, A. (1999). Mathematical Modeling in the Inquiry of Linkages Using LEGO and Graphic Calculator. Does New Technology Alternate Old Technology? In W.-C. Yang, S.-C. Chu, & G, Fitz-Gerald (Eds.), *Proceedings of the Fourth Asian Technology Conference in Mathematics* (pp. 113–122). Asian Technology Conference in Mathematics, Inc.

Liu, J., Zhang, Z., & Liu, Y. (2012). Universal Mechanism Modeling Method in Virtual Assembly Environment. *Chinese Journal of Mechanical Engineering, 25*(6), 1105–1114. https://doi.org/10.3901/CJME.2012.06.1105

Mariotti, M. A. (2002). The Influence of Technological Advances on Students' Mathematics Learning. *Handbook of International Research in Mathematics Education*, 695–723.

Mariotti, M. A., Bussi, M. G. B., Boero, P., Ferri, F., & Garuti, R. (1997). Approaching Geometry Theorems in Contexts. *Pme Xxi, 1*, 180–195. https://doi.org/citeulike-article-id:478750

Martignone, F. (2011). Tasks for Teachers in Mathematics Laboratory Activities: A Case Study. In B. Ubuz (Ed.), *Proceedings of the 35th Conference of the International Group for the Psychology of Mathematics Education* (Vol. 3, pp. 193–200). Middle East Technical University.

Maschietto, M. (2005). The Laboratory of Mathematical Machines of Modena. *Newsletter of the European Mathematical Society*, 57, 34–37.

Maschietto, M., & Bussi, M. G. B. (2011). Mathematical Machines: From History to Mathematics Classroom. In O. Zaslavsky & P.Sullivan (Eds.), *Constructing Knowledge for Teaching Secondary Mathematics. Tasks to Enhance Prospective and Practicing Teacher Learning* (pp. 227–245). Springer. https://doi.org/10.1007/978-0-387-09812-8_14

McCarthy, J. M. (2000). *Geometric Design of Linkages: Vol. 11. Interdisciplinary Applied Mathematics*. Springer-Verlag. https://doi.org/10.1007/978-1-4419-7892-9

McKenna, A. F., & Agogino, A. M. (2004). Supporting Mechanical Reasoning with a Representationally-Rich Learning Environment. *Journal of Engineering Education, 93*(2), 97–104. https://doi.org/10.1002/j.2168-9830.2004.tb00794.x

Taimina, D. (2008, March 5–8). Geometry and Motion Links Mathematics and Engineering in Collections of 19th Century Kinematic Models and Their Digital Representation [Symposium]. In F. Arzarello (Chair), *The First Century of the International Commission on Mathematical Instruction (1908-2008): Reflecting and Shaping the World of Mathematics Education*. ICMI Symposium, Rome, Italy. http://www.unige.ch/math/EnsMath/Rome2008/

Chapter 7

Introducing STEM to Primary Education Students with Arduino and S4A

Panagiotis Michalopoulos[1], Sofia Mpania[1], Anthi Karatrantou[2], & Christos Panagiotakopoulos[2]

[1]*School of Pedagogical and Technological Education (ASPAITE) Branch of Patras*

[2]*Department of Primary Education, University of Patras*

Abstract: *STEM education is aiming to the development of the scientific interest of students and their capability to solve authentic problems, given emphasis to the connection of Science, Technology, Engineering and Mathematics. At the same time simple applications of automatic control systems and robotic constructions are evolving as basic tools of modern life and they provide innovative tools in education as well. In this paper an attempt is made to introduce STEM activities to students of Primary Education supporting them working on simple constructions with Arduino and Scratch for Arduino. Within this framework students were asked to work in groups to design, construct and program their constructions following specially designed worksheets of increasing difficulty with the aim to finally create a Theremin. Analyzing students' work useful information can be derived on how they combined and used in practice knowledge from science, technology and programming and on benefits of educational robotics applications in the frame of STEM education as well.*

Keywords: *STEM Education, Educational Robotics, Arduino in Education, Scratch for Arduino*

1. Introduction

STEM education regards to an integrated scientific approach according to which the four fields of Science, Technology, Engineering and Mathematics, form a whole where elements interact and affect one another. The STEM approach in education requires the use of innovative and alternative methods of teaching and learning, such as projects, laboratory practices and technological tools (Barker et al., 2012; Means et al., 2008). At the same time educational robotics is becoming the next step in education due to its innovative character and the hands-on experience it offers to students making them more receptive to learning stimuli. Educational robotics applications can become a significant educational tool in the STEM approach as students working with robotic constructions are involved in situations that require from them to use and apply knowledge from Mathematics, Science, Technology and Engineering and supported to develop a conceptual basis for the reconstruction of their knowledge (Barker et al., 2012; Eguchi, 2014; Karatrantou & Panagiotakopoulos, 2012).

In this paper an attempt is made to introduce STEM activities to students of Primary Education supporting them working on simple robotic constructions (simple automatic control systems) with Arduino and Scratch for Arduino. Within this framework students were called to work as young makers in an attempt to realize how technology works in real life and asked to work in groups to design, construct and program their simple automatic control systems following specially designed worksheets. Worksheets were of increasing difficulty and supported students to gradually construct and program more complex systems with the aim to finally create a musical instrument, a Theremin.

2. Stem Education and Educational Robotics

STEM education based on the idea of educating students in four disciplines Science, Technology, Engineering and Mathematics in an interdisciplinary and applied approach. Rather than teach the four disciplines as separate and discrete subjects, STEM integrates them into a cohesive learning approach based on real-world applications and situations. The objectives of STEM education, at all levels of education, include the development of the scientific interest of students and their capability to solve authentic problems aiming knowledge from Science to be used for the understanding of the natural world around. Moreover, students to be able to use new technological tools and understand how Technology affects the surrounding world, to realize the significance of Engineering in real world and how they are linked each other. It is also aiming students' abilities related to Mathematics, such as analysis, documentation, and problem solving to be improved, supporting them to cope with situations in their everyday lives (National Research Council, 2011; Means et al., 2008; Thornburg, 2008). In his framework STEM education serves efforts to increase students' interest in pursuing Science, Technology, Engineering and Mathematics Studies and Careers. For the success of STEM education programs, however, certain parameters have to be taken into consideration. These include a demanding content, a research learning environment, pre-defined educational outcomes, clear objectives, innovative educational scenarios and commitment and support from society as well (Bayer Corporation, 2016; Stergiopoulou et. al, 2016).

On the other hand, robotics since the late 80s has been used in almost all levels of education as an auxiliary tool in teaching various concepts of subjects such as Mathematics, Sciences, Engineering, Technology and Computer science and as a teaching subject as well (Karatrantou & Panagiotakopoulos, 2012; Mubin et al., 2013). Robotic constructions are used in education as extracurricular activities and as educational activities aligned to the curriculum objectives supporting the development of the 21st century skills such as collaboration, problem-solving, creativity, critical thinking and computational thinking. Educational robotics is a dynamic and strong promising tool for STEM education. It is offering to students 'objects to think with', as Papert (1993) described supporting them to realize how technology works in real life, to see computing in a different way, to cultivate computational thinking. It is offering to teachers strategies and tools to manage a shift from 'Black box' approaches in education (where students act as consumers) to 'transparent box' approaches (where students can work as constructors).

Many educational robotics kits are used in all educational levels such as Lego Mindstorms (RCX, NXT, EV3), BoeBot, Activity Bot, Arduino. In this research the Arduino platform was used for the construction of the simple systems and the software Scratch for Arduino (S4A) was used for the programming and control of the systems.

3. Arduino

Arduino is a low-cost open-source electronics platform based on easy-to-use hardware. It is intended for anyone who wants to work on interactive projects, demanding basic programming knowledge and experience and minimum knowledge on electronics. Arduino 'senses' the environment by receiving inputs from many sensors, and '*affects*' it by controlling lights, motors, and other actuators. It was created at the Ivrea Interaction Design Institute as an easy tool to be used by students without a background in electronics and programming. Nowadays Arduino has been the heart of projects, from everyday objects to complex scientific instruments. A worldwide community of students, hobbyists, artists, programmers and professionals has gathered around it and support novices and experts as well (https://www.arduino.cc/). During the last years Arduino is used in schools at all levels of education in many countries to support teaching and learning in science (physics, chemistry, biology), technology, computer science, supporting experiments in laboratories and by distance as well (Fotopoulos et. al., 2016; Orfanakis and Papadakis, 2014; Rubio et al., 2013).

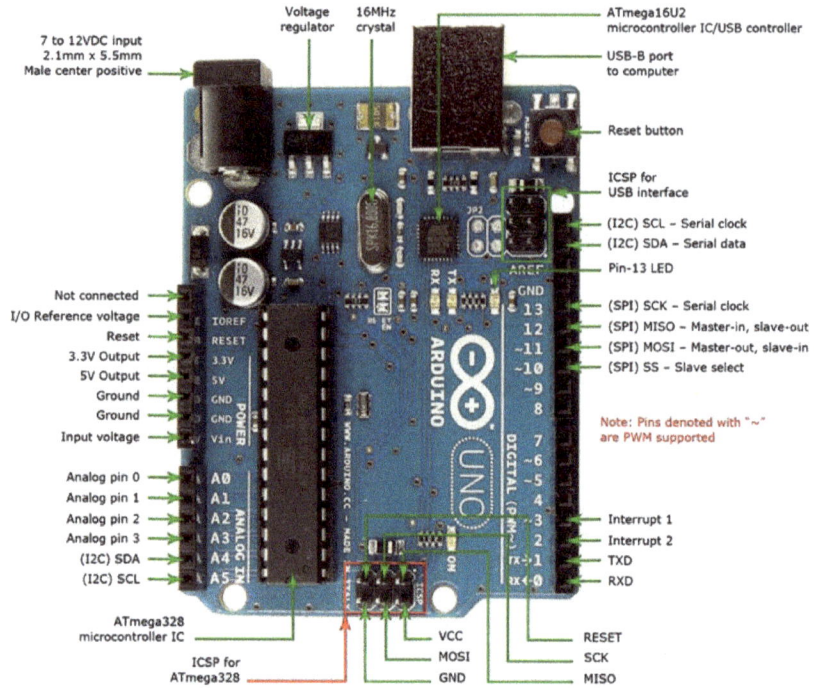

4. Scratch for Arduino

Scratch for Arduino (S4A) is a Scratch modification that allows users to program the Arduino platform. It includes new blocks for managing sensors and actuators connected to Arduino. The main aim is to attract young people to the programming world and to provide them a friendly interface interacting with a set of boards through user events (http://s4a.cat/).

5. Methodology

The aim of the study was to introduce STEM activities to students of Primary Education supporting them working on simple automatic control systems with Arduino and Scratch for Arduino and to analyze their work in order to derive useful information on how they combined and used in practice knowledge from science, technology and programming. Additionally, useful data on the benefits of educational robotics applications in the frame of STEM education can be extracted.

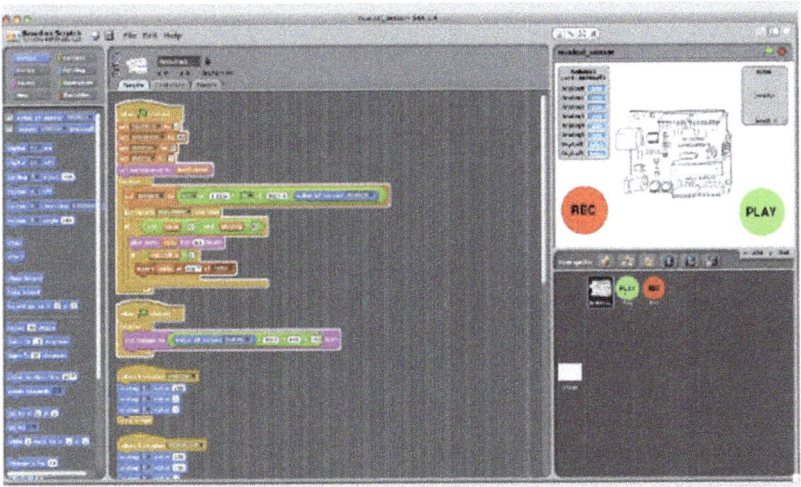

Figure 2: Scratch for Arduino: a typical screen shot

Students while working on the specific worksheets with the aim to gradually construct a Theremin were expected to recall and make use of knowledge from their science subjects, be familiar with and recognize basic electronic circuit elements, describe how a circuit works, be able to explain how a simple control system works, use basic programming structures, choose the appropriate programming structures to control their construction, realize that a connection among subjects of Engineering, Science, Mathematics and Technology exists, recognize applications of automatic control systems in everyday life, adopt positive attitude for innovation in everyday life.

The research took place in a public Primary School in Patras. A total of eighteen (18) sixth grade students participated voluntarily in the study (eight (8) girls and ten (10) boys). The students were divided randomly into five (5) groups of three persons (2 groups) and four persons (3 groups) and the educational activity took place during three sessions in the Computers Laboratory of the schools (2 two-hour sessions and 1 one-hour session). Students of the sixth grade have coped with the widest range of subjects and have been taught about electricity, circuits and their elements (sources, lamps, switches and resistors), light and their basic properties. They are familiar with the Scratch programming environment and have used basic programming structures in simple programs.

The whole activity was based on Worksheets of increasing difficulty that supported students to gradually construct and program more complex systems with the aim to finally create a musical instrument, a Theremin. The 1st session started with a Diagnostic Questionnaire and a short discussion on the topic of robotic and automatic control systems. A demonstration of a Theremin made with Arduino helped them to understand what they should finally make and worked as a challenge. A familiarization phase with Arduino platform and S4A took place based on the 1st worksheet according to which students worked on a simple circuit with a LED and a resistor and discussed about the program to make it work. During the 2nd session, the 2nd and the 3rd worksheet was the basis for students' work. According to the 2nd worksheet students had to add a new element on the circuit, a switch to control the function of the circuit and modify their program using the control structure 'IF…THEN… ELSE…' in order for the circuit to work properly. The 3rd worksheet supported students to construct their Theremin using a photo-sensor (a photo-resistor) on the circuit and a variable in their S4A program. During the 3rd session a discussion about the whole activity and what they worked on, a discussion concerning applications of simple automatic control systems in everyday life and the evaluation of the whole procedure (Evaluation questionnaire) closed the procedure.

For the purposes of the study four data collection methods were used: monitoring (groups and members) and personal notes of the researchers, recording of groups' discussions, analyzing answered worksheets by the students while working, answering of short questionnaires before (Diagnostic Questionnaire) and after the whole activity (Evaluation Questionnaire). Three teachers-researchers were observing the discussions, activities and reactions among the students. They kept notes and made interventions when students needed help, adopting a supportive and facilitative role of students' work and learning. The Diagnostic Questionnaire consisted of seven (7) closed-type questions and students had to give anonymous answers about their gender, their previous programming experience, their interest in learning more in Programming and what they mainly used the computer for. The Evaluation Questionnaire was used to evaluate the overall procedure (the robotic kit as a tool, the programming environment, the learning outcomes) and consisted of seven (7) closed- type and six (6) open-ended questions.

6. Students' Work and Findings

According to the students' answers to the diagnostic questionnaire, 17 students (94%) usually work a lot with computers (most of them for fun, for searching information needed for school projects, for communicating with friends) but none of them had an experience in programming more than what they programmed during school lessons. The majority of the students (17 students, 94%) think that programming is a difficult task. The demonstration of the Theremin at the beginning triggered their curiosity and their enthusiasm and their willing to start working immediately.

Starting working students were asked to draw an electric circuit according to what they already knew from their physics lessons aiming to recall necessary knowledge in order to discuss about the electrical elements and circuits they should construct later. All students drew correctly their circuits and only one group of students put the polarity of the source in a wrong direction. According to the 1st worksheet, students had to create a circuit on the breadboard based on an already designed circuit. All students recognized the resistor and the LED, a discussion took place on the role of the resistor in the circuit and in a very short period of time all groups of students created their circuits correctly. A program in S4A, that should turn on the LED for some seconds and after that should turn off the LED for some seconds, was given to them in order to describe and test its result on the circuit. Students had to modify the program (using the repeat structure without a condition the forever block) in order to make the LED blinking continuously with a rate of their choice and all of them succeed it.

According to the 2nd worksheet students had to add a switch in the circuit. They drew the new circuit correctly and created it correctly too based on the already designed circuit. A program in S4A, which should turn on the LED when the switch was pressed and should turn off the LED when the switch was released, was given to them in order to describe and test its result on the circuit. A discussion about the role of the open-switch and the closed-switch as well as about the role of the control structure 'IF...THEN...ELSE...' took place. Most students faced difficulties with the control structure.

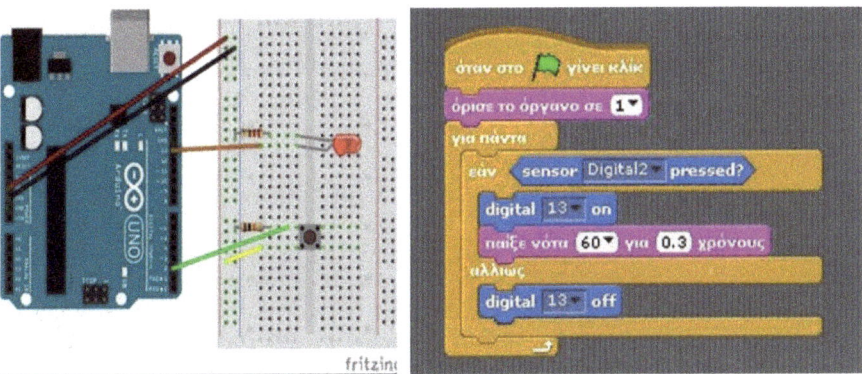

Figure 3: The circuit and the program for Worksheet 2

According to the 3rd worksheet students had to use a photo-sensor (a photo-resistor) to create and program a Theremin. They created the circuit based on the given design and made the program in S4A according the given one. They tested the result of the program on the circuit and a discussion took place on how the photo-sensor works, what it 'senses' and how, how it affects the circuit, what are the roles of the commands in the program. Many of the students faced difficulties to understand how the photo-sensor works. Students' ideas on what a photo-sensor is were intuitional and based on their initial impression when they saw the

Theremin at the beginning of the activity: '...*it can realize the light above it...*', '...*it can understand when we have our hand over it...and how close...*', '...*when it is darker around the sound is heavy and when its lighter around the sound is higher...*'. Students introduced to the concept of 'variable' in programming to keep the characteristic value of the musical note the Theremin should play. Most students faced difficulties to understand what a 'variable' in programming is and what its role is.

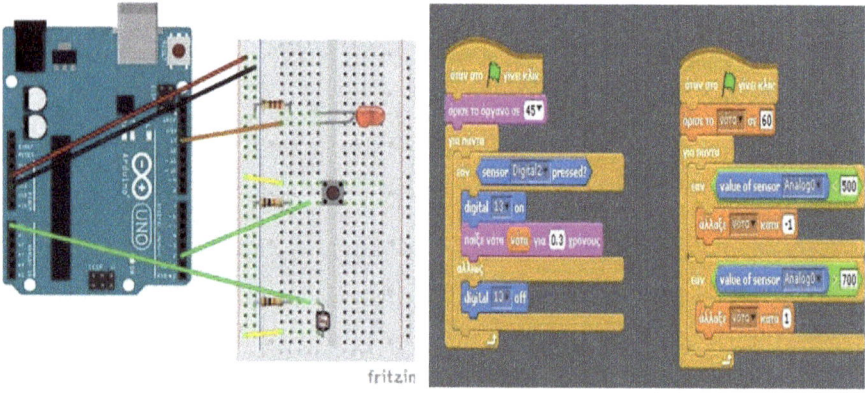

Figure 4: The circuit and the program for Worksheet 3

Students' answers at the assessment questionnaire showed that all of them liked the activity much and very much. Fourteen (14) of them (77.7%) would like to work on such constructions and applications again. Thirteen (13) students (72.2%) said they didn't face any difficulty, 3 students (16.7%) faced difficulties in programming and 2 students (11.1%) during creating the circuit. Most of them answered that '...*they did not face difficulties because we were a team and each one made something to complete the circuit and the program...*'. Answers concerning the programming structures used showed that all students seemed to understand and were able to describe how the repetition structure works and when it is used but only 5 of them (27.8%) could do the same for the control structure 'IF...THEN...ELSE...'. Most of the students gave descriptive but not accurate answers. Most students faced difficulties with the 'variable' in programming.

Answers concerning the elements of the circuits showed that all students could recognize all the elements used. Thirteen (13) of them (72.2%) could describe their properties and the role in the circuits they created, 4 students (22.2%) could recognize only the switch and the LED and only one (1) student could not give correct answers. Typical answers on the function of the elements in the circuit were: '...*when we pressed the switch the LED was on the cables are connecting the elements of the circuit ...the resistor is decreasing the current in the circuit...*', '...*the resistor is decreasing the current. The cables are connecting different things and the switch let the current pass into the circuit...'.* It was difficult for them to describe the function of the photo-sensor. Most students could propose a new application: *alarms, a sensor at home to open doors and windows depending on light or temperature, automatic watering of plants, automatic faucet to wash our hands....*

Figure 5: Students working

It is important to notice that the members of all groups worked in cooperation. Each member tried to contribute to the activity, there was a work and responsibilities sharing among members and leaders were appeared in most groups. In all groups discussion among members took place to find solutions and complete their task each time. Three groups asked for help from the teacher while working.

7. Discussion and Conclusions

In this paper an attempt is made to introduce STEM activities to students of Primary School supporting them working on simple robotic constructions with Arduino and Scratch for Arduino. Students are asked to work in groups to design, construct and program simple automatic control systems following three specially designed worksheets of increasing difficulty and guided students to gradually construct and program more complex systems with the aim to finally create a Theremin. Students worked during three sessions (5 didactic hours) on the activity in an attempt to be supported to realize how technology works in real life.

According to the data derived from the monitoring of students' work (groups and members), the personal notes of the researchers, the recording of groups' discussions, the analysis of the answered worksheets by the students and their answers on the short evaluation questionnaire useful information derived. All students worked effectively as members of their group. All students seemed to recall knowledge from physics about circuits and light, seemed to connect this knowledge with their work during this activity, designed and created circuits correctly. Most of them recognized the elements of the circuits and could describe correctly their properties and their role in the circuit. Most of them faced difficulties with the photo-sensor. All of them understood the role of the repetition structure in programs but some of them faced difficulties with the structure 'IF...THEN...ELSE...' and the concept of 'variable'. Most students could propose a new application using automatic control systems in everyday life and expressed their interest to work more on such activities.

Concluding, such kind of educational activities could be dynamic tools in the frame of STEM education and could raise student's interest and engagement in STEM through the use of simple robotic constructions on the basis of well-structured educational scenarios according to the needs of students each time.

Closing, it should be emphasized that improving the interest and motivation of students for STEM education is a complex issue. A range of new approaches need to be implemented and examples of good practice need to be mainstreamed. Since the Lisbon agenda was launched by the European Council in 2000, a lot of attention has been focused on Europe's need to foster a dynamic and innovative knowledge-based economy, not least by producing an adequate output of scientific specialists (Joyce and Dzoga 2012). To achieve this goal we need to increase participation in Science, Technology, Engineering and Mathematics (STEM) studies and career. That means that we need teachers training on the field as well as suitable educational approaches, scenarios, material and practice.

REFERENCES

Barker, B. S., Nugent G., Grandgenett, N., & Adamchuk, V. I. (Eds.). (2012). *Robots in K-12 Education: A New Technology for Learning*. IGI Global.

Bayer Corporation (2016). *Planting the Seeds for a Diverse U.S. STEM Pipeline: A Compendium of Best Practice K-12 STEM Education Programs*. MSMS Resources. https://www.makingsciencemakesense.com/static/documents/Resources/K-12-STEM-edu-programs.pdf

Eguchi, A. (2014). Educational Robotics for Promoting 21st Century Skills. *Journal of Automation, Mobile Robotics & Intelligent Systems, 8*(1), 5–11. https://doi.org/10.14313/jamris_1-2014/1

Fotopoulos, V, Spiliopoulos, A. I., & Fanariotis, A. (2016). Preparing a Remote Conducted Course for Microcontrollers Based on Arduino. In A. Lionarakis (Ed.), *Proceedings of the 7th International Conference in Open & Distance Learning* (Vol. 5, Pt. 2, pp 133–139). https://doi.org/10.12681/icodl.563

Joyce, A., & Dzoga, M. (2012). *Intel White Paper: Science, technology, engineering and mathematics education overcoming challenges in Europe*. Intel Educator Academy EMEA.

Karatrantou, A., & Panagiotakopoulos, C. (2012). Educational Robotics and Teaching Introductory Programming within an Interdisciplinary Framework. In A. Jimoyiannis (Ed.), *Research on E-Learning and ICT in Education* (pp. 197–210). Springer.

Means, B., Confrey, J., House, A., & Bhanot, R. (2008, October 15). Specialized Science Technology Engineering and Mathematics Secondary Schools in the U.S. *STEM High Schools | SRI International*. Retrieved March 17, 2018, from https://www.sri.com/sites/default/files/publications/imports/STEM_Report1_bm08.pdf

Mubin, O., Stevens, C. J. Shahid, S., Al Mahmud, A., & Dong, J.-J. (2013). A Review of the Applicability of Robots in Education. *Technology for Education and Learning, 1*(1), 1–7. https://doi.org/journal.209.2013.1.209-0015

National Research Council (2011). *Successful K-12 STEM Education: Identifying Effective Approaches in Science, Technology, Engineering, and Mathematics*. The National Academies Press. https://doi.org/10.17226/13158

Orfanakis, V., & Papadakis, S. (2014). Μια δραστηριότητα διδασκαλίας προγραμματισμού με τη χρήση του Scratch για Arduino (S4A) [An Educational Activity for Programming Using Arduino and Scratch for Arduino (S4A)]. In N. Alexandris, C. Douligeris, P. Vlamos, & Belesiotis, V. (Eds.), *Proceedings of the 6th Conference on Informatics in Education CIE2014* (pp. 540–549). GREEK COMPUTER SOCIETY (GCS)

Papert, S. (1993). *The Children's Machine: Rethinking School in the Age of the Computer*. BasicBooks.

Rubio, M. A., Hierro, C. M., & Pablo, A. P. D. M. (2013). Using Arduino to Enhance Computer Programming Courses in Science and Engineering. In L. G. Chova, A. L. Martínez, I. C. Torres (Eds.), *EduLearn 13 Proceedings: 5th International Conference on Education and New Learning Technologies* (pp. 5127–5133). IATED.

Stergiopoulou, M., Karatrantou, A., & Panagiotakopoulos, C. (2016). *Educational Robotics and STEM Education in Primary Education: A Pilot Study Using the H&S Electronic Systems Platform*. Educational Robotics in the Makers Era (Edurobotics 2016). Springer International Publishing.

Thornburg, D. D. (2008). Why STEM Topics Are Interrelated: The Importance of Interdisciplinary Studies in K-12 Education. *Thornburg Center for Space Exploration and Ardusat*. Thornburg Center for Space Exploration. Retrieved March 17, 2018, from http://tcse-k12.org/pages/stem.pdf

Chapter 8

CREATIONS: Developing an Engaging Science Classroom

Ioannis Alexopoulos[1], Sofoklis Sotiriou[1], Zacharoula Smyrnaiou[2], Menelaos Sotiriou[2], & Franz Bogner[3]

[1]*Ellinogermaniki Agogi, Greece*

[2]*Faculty of Philosophy, Pedagogy and Psychology, National and Kapodistrian University of Athens, Greece*

[3]*Didactics of Biology/ Centre of Math & Science Education, University of Bayreuth, Germany*

Abstract: *Taking into account the strongly decreased interest of young people in science and mathematics, this study aims to propose a new creative pedagogic approach in order for this tendency to be reversed, developed in the framework of EU project CREATIONS. The first implementation activities of this approach, addressing both students and teachers, have been already taken place and the initial results of the implementation leads to quite positive conclusions, concerning the motivation and interest in learning science.*

Keywords: *Creativity, Inquiry Based Science Education, Art, Large Research Infrastructure*

1. Introduction

The publication of the "Science Education Now: A renewed Pedagogy for the Future of Europe" report brought science and mathematics education to the top of educational goals of the member states (Rocard et al., 2007). The authors argue that school science teaching needs to become more engaging, based on inquiry based and problem-solving methods and designed to meet the interests of young people. According to the report, the origins of the alarming decline in young people's interest for key science studies and mathematics can be found, among other causes, in the old-fashioned way science is taught at schools. The crucial role that positive contacts with science at a younger age have in the subsequent formation of attitudes toward science has been emphasized in many studies (OECD, 2014a.). However, traditional formal science education too often fails to foster these, affecting thus negatively the development of adolescents' attitudes towards learning science. Also, the tension created between objectivism (the objective teacher-centered pedagogy) and constructivism (the constructive and student-centered pedagogy) represents a crucial classroom issue influencing teaching and learning (Kinchin, 2004). The TIMSS (Third International Mathematics and Science Study) 2003 International Science Report specifically documented that the three activities accounting for 57 percent of class time were: teacher lecture (24%), teacher-guided student practice (19%), and students working on problems on their own (14%) in science classes in the European countries participating in the study (Martin et al., 2004). Furthermore the recent TALIS (Teaching and Learning International Study) results demonstrate that the current science classroom learning environment is dominated by traditional pedagogies that are not able to support the introduction of the scientific methodology (OECD, 2014b). The fact is that there is a major mismatch between opportunity and action in most education systems today. This revolves around the meaning of "science education," a term that is often misappropriated in the current school practice, where rather than learning how to think scientifically, students are generally being told about science and asked to remember facts (Alberts, 2009).

This disturbing situation must be corrected if science education is to have any hope of taking its proper place as an essential part of the education of students everywhere. However, school practices have not changed in ways that reflect this progress. Moreover, modern technologies (e.g., use of social networking tools, remote and virtual labs, advanced visualizations, simulations, virtual worlds and shared collaborative environments), which go beyond the use of simple applications and the internet have not been fully integrated/incorporated in the current science learning environment. According to the recent work performed in the framework of the large-scale initiative PATHWAY the deeper problem in science education is one of fundamental purpose (Sotiriou & Bogner, 2011). Schools, the authors argue, have never provided a satisfactory education in sciences for the majority. Now the evidence is that it is failing even in its original purpose, to provide a route into science for future scientists. The challenge therefore, is to re-imagine science education: to consider how it can be made fit for the modern world and how it can meet the needs of all students; those who will go on to work in scientific and technical subjects, and those who will not.

In this framework, CREATIONS (http://creations-project.eu/) as a multinational EU project is aiming to demonstrate innovative approaches that involve teachers and students in Scientific Research through creative ways (from STEM to STEAM). By basing on Arts and focusing on effective links and synergies between schools and research infrastructures, young people's interest in science and in following scientific careers is expected to be affected. The project is addressing the potential impacts of international research facilities on advancing science education, in using case studies from one of the largest research infrastructures of the world: the European Organization for Nuclear Research (CERN).

1.1. Concept and Approach

In our study a creative approach in science education is presented, in order to generate alternative ideas and strategies within scientific enquiry as an individual or community. Figure 1 offers an overview on this pedagogic approach. At the core of the proposed approach are the creative scenarios and school-based activities and the accompanying pedagogic principles. **Creative science education** is the main context within which the project is developed. At the bottom of the graph **arts education philosophy and methods** is positioned as a 'holder' within which creative science education is being nurtured, grown or 'encultured' via arts practice. As we move in towards the center of the graph, we can see that one of the main drivers for CREATIONS creativity is possibility thinking for all involved. This means being able to ask 'what if' and 'as if' questions such as:

- What if I/we choose to explore this scientific question rather than that one?

- What if I/we use this arts approach to help me explore my question?

As we move in another layer towards the center of the graph, we can see four key defining features of engaging science classroom environments. These are the 4Ps of engagement in creative science education (Craft, 2011):

- **pluralities:** opportunities for students and teachers to experiment with many different places, activities, personal identities, and people

- **possibilities:** opportunities for possibility thinking, transitioning from what is to what might be, in open possibility spaces

- **participation:** opportunities for students and teachers to take action, make themselves visible on their own terms, and act as agents of change

- **playfulness:** opportunities for students and teachers to learn, create and self-create in emotionally rich, learning environments.

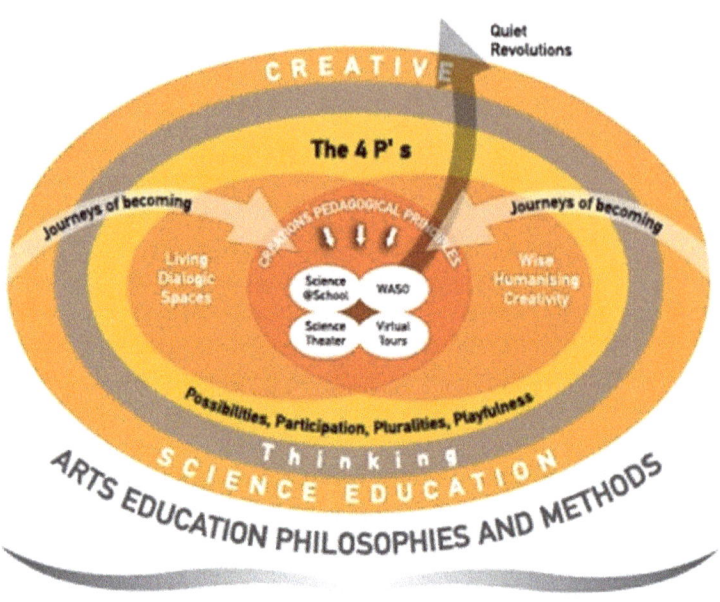

Figure 1: CREATIONS Approach: The graph demonstrates how the CREATIONS scenarios support the development of a research culture in science classrooms.

We then come closer to the heart of the CREATIONS graph and find WHC (wise humanizing creativity) and LDS (living dialogic space). The WHC that is being sought in CREATIONS is not only an individual activity, but also happens in collaboration with fellow learners, teachers and other adult professionals (artists, researchers). These individual and collaborative creative activities form part of a wider web of ethically-guided communal interaction geared towards both helping children and young people become more creative scientists and assisting teachers in becoming more creative in how they teach science. For this reason WHC is positioned very close to the heart of the CREATIONS graph as it is one of the core aims of the CREATIONS pedagogic principles. Alongside and integrated with WHC, is LDS, always a partner to WHC in terms of conceptualizing ideas and developing practice. Again LDS is at the heart of the CREATIONS graph because its methods (participation, emancipation, working bottom up, debate and difference, openness to action, partiality, and acknowledging embodied and verbal modes of knowing) are fundamental to allowing WHC to happen. Chappell et al.

(2012) have evidenced the importance of dialogue at the heart of engaged, creative learning in the arts and it is this kind of dialogue that has been highlighted and applied within the CREATIONS approach. Via these processes the aim is to develop creative young scientists and creative science teaching pedagogies. Embedded within this is the vitally important notion that students and teachers are creating wisely and humanely, and that cyclical developments occur between their creativity and their identity. As they generate new ideas; this in turn generates change in them as 'makers'; they are also developing or 'becoming' themselves. Slowly, small changes accumulate to contribute to '**journeys of becoming**' (shown developing across the layers in Figure 1). These individual journeys accumulate together, embedded within an ethical awareness of the impact of creative actions on the group. Through this process small-scale creative changes or '**quiet revolutions**' can take place for the group as a whole (shown as emerging from the heart of the *CREATIONS* activity).

1.2. CREATIONS Demonstrators

The implementation of the described pedagogic approach highlights and promotes best practices in introducing scientific work in science classrooms. The aim was to offer to the teachers, who will be involved in the project activities, a variety of resources that is arranged so that it does not impose a fixed curriculum, but instead supports the development of a model that can be customized to reflect location, culture and ideology. These initiatives are implemented the last years in CERN, but also in numerous schools in Europe and they have proven their efficiency as practices that introduce the scientific methodology in the science classroom.

In the framework of the project we enrich these initiatives with the proposed creative approach in order to increase the utility of them through coordination, systematic dissemination and effective teachers' community building. The **CREATIONS Demonstrators** that emerged are treated as case and will be disseminated in different environments (teachers' preparation and professional development institutions, schools, science and research centers) across Europe during the life cycle of the project, in order for this pedagogic approach to be tested. The process of observing and reflecting on teachers actions, and on students' learning and thinking, can lead to changes in the knowledge, beliefs, attitudes, and ultimately the school everyday practice. A short description of demonstrators which lying on the core of the CREATIONS pedagogic approach follows:

> **Art@CMS** is an education and outreach initiative of the CMS experiment at CERN that seeks to act as an inspiring springboard for engaging the public in general, and youth in particular, in the excitement of scientific research in High Energy Physics. In 2014 Art@CMS events and workshops have been taken place in 7 countries involving more than 700 students.

CERN virtual visits. CERN in cooperation with the European initiative **Open Discover Space** (portal.opendiscoveryspace.eu) is offering the opportunity to school students to perform a virtual field trip to CERN experiments. Students virtually guided through the research infrastructure, communicate with scientists in their mother language, ask questions, and learn about the research work at CERN. More than 50 virtual visits were organized in 2014 involving more than 10,000 students from different European countries.

HYPATIA analysis tool enables high schools students together with their teachers to study the fundamental particles of matter and their interactions, through examining the graphic visualization/display of the products of particle collisions at LHC world's most powerful particle accelerator. These products are "events" detected by the ATLAS experiment.

Write a Science Opera (WASO) is a creative approach to inquiry-based music and science education in which students of different ages, supported by teachers, opera artists and scientists are the creators of an educational performance. The WASO concept was developed at Stord/Haugesund University College (Norway) as well as the Royal Opera House (London)'s Education department.

Taking into consideration the described framework in this study we investigate:

Can creative teaching scenarios, such as CREATIONS Demonstrators, improve the motivation and interest in learning science?

2. Setting of the Study

A common framework was created for the design and development of a series of demonstrators that introduce effectively scientific methodology and culture in science classrooms. For each one of these demonstrators an on line created in the Open discovery space platform (http://www.opendiscoveryspace.eu/), in order to support users of the demonstrators. Moreover it is requested for every new demonstrator, an on line support community to be developed. Such communities act as a context of cooperation within and between schools, universities, research institutions, artists and encourage development and evaluation of instruction, exchange of ideas and best practices, providing at the same time support and stimulation from research.

Figure 2: The map of the support online communities of the initial CREATIONS Demonstrators

The CREATIONS Project includes large-scale pilots of a variety of activities to be implemented in local, national or international level in numerous countries.

2.1. Implementation Activities

The first implementation activities witch based on a set of different CREATIONS demonstrators are the 6-days teachers international training course held in Marathonas and the students 5-days national summer school held in Messini Greece, both during July.

In teachers training Course, 17 teachers took part from from 6 European countries (Greece, Estonia, Sweden, Croatia, Finland & Switzerland). The course aimed to present to the teachers innovative approaches and activities that involve students in Scientific Research through creative ways that are based on Art. The course included lectures, workshops and activities concerning: The CREATIONS pedagogical framework, creative approaches to teach and communicate science, the science concept of neutrino particle detection particles which had to be communicated via the Science Operatic performance, composition/use of music in a learning activity, how to write a Science opera, the management of a large-scale learning activity involving art and science, etc.

Concerning the case of students summer school, 50 high school students participated from all over Greece. Main aim of the summer school was to introduce the students in complex science topics such as universe evolution, Higgs boson, gravitational waves, neutrino telescopes and in the same time by using art enhance the creative thinking of students. To this end the students visited NESTOR institute at Pylos, as well as make a virtual visit to ATLAS experiment at CERN. Furthermore they conducted lab exercises which simulated the detection techniques of elementary particles etc. Participate in science café. Moreover they learn to use creative approaches in order to communicate science concept. All this acquired knowledge from the activities mentioned above, was used by the students in order to design 5 science stories and perform relevant short plays.

2.2. Data Collection and Analysis

2.2.1. Teachers Training Course

A quantitative approach with questionnaires was implemented, as well as qualitative approach with interviews was realized. Specifically 15 out of 17 teachers completed a pre-test before participation and a post-test afterwards. The questionnaires included questions concerning demographic data, participants' science motivation and technology interest. More over 14 participants were interviewed about their expectations of the training course and their personality as a teacher during the course.

2.2.2. Students Summer School

The methodology employed to analyze scientific data gathered from the theatrical performances in the summer programme of Messini, constitutes a merging of qualitative and quantitative analysis (10). The data were analyzed and classified into categories. This categorization took into consideration the theoretical framework of the analysis along the empirical evidence gathered from the theatrical plays performed by students. Student representation of scientific concept and the production of scientific meaning was studied using 3 categories.

- Embodied Learning
- Multiple representational systems (verbal, embodied, digital, kinesthetic representation, elements of Art)
- Analogical Reasoning

Each category is further divided into subcategories/ properties which are connected to basic features of embodied learning, of multiple systems of symbols, of analogical reasoning.

3. Initial Results and Discussion

Preliminary findings based on a data analysis from Creations 6-days teachers training course show that the creative approach of the course and the new ways of teaching and learning were inspiring for them and increase their interest in learning science. Furthermore the participants stated that they want to take part again in training like this, in order to learn new innovative learning methods and different demonstrators.

Concerning the findings from the students' 5-days summer school, we can reach the conclusion that students' interest and motivation in science was quite increased after the summer school. Indicatively, one of Student that participated in the course a student who completed the first year of senior high school stated: "My participation in the summer school was one of the best experiences in my life!". Students employed scientific concepts in all of the plays. As far as the representation of scientific concept and the creation of meaning are concerned, students seemed understand all sub-elements and basic characteristics of each concept. They managed to render the general meaning of the concepts and to explain simple scientific terminology. It is significant to mention that students were able to use simple language to explain scientific terminology at the same time they were using this terminology provided they had understood the scientific concept in question. In most cases, they used simple everyday objects, which verify that they gained, built and appropriated knowledge. This means that they managed to successfully connect newly gained knowledge with everyday life and to use it in an everyday environment

Furthermore both training courses seems to managed to combine teaching about Science with creativity and Art, offering a new way of approaching these cognitive areas. Overall the first findings from the existing data and the positive reactions so far of the students and teachers involved, allow us to estimate that the whole process will enhance students' motivation in science and also their creativity. We will be able to provide more details on this subject mutter during the conference.

Acknowledgments

This study has been developed in the framework of the HORIZON 2020 project CREATIONS (Grant Agreement 665917).

REFERENCES

Alberts, B. 2009. Redefining Science Education. *Science,* *323*(5913), 437. https://doi.org/10.1126/science.1170933

Chappell, K., Craft, A., Rolfe, L., & Jobbins, V. (2012). Humanising Creativity: valuing our journeys of becoming. *International Journal of Education and the Arts, 13*(8), 1–35.

Craft, A. (2011). *Creativity and Educational Futures.* Trentham Books.

Gobo, G. (2005). The Renaissance of Qualitative Methods [22 paragraphs]. *Forum Qualitative Sozialforschung / Forum: Qualitative Social Research, 6*(3). http://dx.doi.org/10.17169/fqs-6.3.5

Kinchin, M. I. (2004). Investigating students' beliefs about their preferred role as learners. *Educational Research, 46*(3), 301–312. https://doi.org/10.1080/001318804200277359

Martin, O. Michael, I. V. S. Mullis, Gonzalez E. J., & Chrostowski, S. J. (2004). *TIMSS 2003 International science report*. TIMSS & PIRLS International Study Center, Lynch School of Education, Boston College.

OECD (2014a). Italy. In OECD (Ed.), *PISA 2012 Results: What Students Know and Can Do: Vol. I. Student Performance in Mathematics, Reading and Science* (Rev. ed). OECD Publishing.

OECD (2014b). *Talis 2013 Results: An International Perspective on Teaching and Learning*. OECD Publishing. http://dx.doi.org/10.1787/9789264196261-en

Rocard, M., Csermely, P., Jorde, D., Lenzen, D., Walberg-Henriksson H., Hemmo, V. (2007). *Science Education NOW: A Renewed Pedagogy for the Future of Europe*. Office for Official Publications of the European Communities

Sotiriou, S., & Bogner, F. (2011). Inspiring Science Learning: Designing the Science Classroom of the Future. *Advanced Science Letters*, 4(11–12), 3304–3309. https://doi.org/10.1166/asl.2011.2039

Chapter 9

Tangible User Interfaces in Early Year Mathematical Education: An Experimental Study

Rizos Chaliampalias[1], Anna Chronaki[2], & Achilles Kameas[3]

[1]*Secondary Education*

[2]*Department of Early Childhood Education, University of Thessaly*

[3]*School of Science and Technology, Hellenic Open University*

Abstract: *The use of metaphors is essential in the learning design of tangible user interfaces. Especially for mathematics learning the use of metaphors is paramount for the development of number sense in children. Our research discusses how children can identify multiple correct solutions to numerical problems, and specifically additive structure problems, when conceptual metaphors are being implemented as tangible interfaces, compared to situations where no use of metaphors is made. In order to evaluate the potential for tangible interfaces to support children's numeracy skills, it is important to first identify the possible advantages, as well as limitations, of using conceptual metaphors in this domain.*

Keywords: *tangible user interfaces, conceptual metaphors, physical manipulatives*

1. Introduction

During the last two decades different approaches of human-computer interaction have emerged. In these new forms of interaction we are being asked to move beyond the use of mouse and keyboard towards the use of multi-touch, augmented and virtual reality technologies. In addition to graphical user interfaces (GUIs), another type of interface, tangible user interfaces (TUIs) is being developed. In this interface, the mapping between the manipulation of the input device and the resulting digital output is tightly coupled. In tangible user interfaces, everyday objects are embedded with computational power and links with corresponding digital representations of concrete information are enabled. The aim of our study is focused towards exploring the role of TUIs in problem solving by comparing young children's ability to solve addition problems using physical manipulatives, with their ability to solve such problems using equivalent digital representations. In the present paper we examine the hypothesis that the mapping movement of physical manipulatives to digital representations through the conceptual metaphor of pan balance will enhance children's performance in additive structure problems. Specifically, the aims and objectives of our research were as follows: first, to design and implement a TUI system applied to mathematics learning skills for children in the first years of primary school and second, to investigate the role of conceptual metaphors in the TUI system and during its situated use for addition type mathematical problems.

2. Background

The distinction between physical and digital becomes increasingly blurred. Multiple representations can be designed for TUIs in order to reduce extraneous cognitive computation by using one digital representation to support interpretation of an abstract concept using physical objects. Interaction mappings could be based on image schema by using conceptual metaphors. Mappings between properties of physical artifacts and corresponding digital objects can be structured upon the metaphorical relations of image schemata. Our main research focus was the evaluation of the potential for tangible technologies to support children's mathematical problem solving skills in relation to number. Specifically, our study examined the role of multiple representations and conceptual metaphors in improving children's ability for solving addition problems. This was addressed by conducting a between subjects' design with condition conceptual metaphor (No/Yes) as the between subjects independent variable. The children sat in front of a table that contained red cubes. Children in the second group were also sat at the same table with cubes as the first group. However, in the second group, the manipulation of cubes by the children triggered an update on an adjacent computer screen that showed the moved cubes being placed on a pan. The children were asked what he/she would have to do so that the frog could have as many cubes as the turtle. Analysis of the experimental results has shown that children can produce more correct solutions when a metaphor such as the pan is introduced in the problem description. Our findings reinforce the hypothesis that in order to evaluate the potential for tangible technologies to support children's numerical development, it is important to first identify suitable conceptual metaphors in this domain.

3. Learning with Tangible Technologies

TUIs show a potential to enhance the way in which people interact with digital information using physical artifacts. TUIs are one of the concepts that Dourish (2001) has developed under the term tangible computing. The other forms of tangible computing are Ubiquitous Computing, Augmented Reality, Reactive Rooms and Context-Aware Devices. TUIs attempt to take everyday objects and invest them with computation instead of using a keyboard and a mouse to point and click on virtual objects. Ullmer and Ishii (2001) presented the first definition of tangible user interfaces by providing a framework in which they identify three key characteristics of physical and digital relationships: physical representations are coupled with digital information, embody control of digital representations while they are perceptually coupled with mediated digital representations. Tangible user interfaces (TUIs) provide a much closer coupling between the physical and digital – to the extent that the distinction between input and output is eliminated. Koleva et al. (2003) distinguish between different types of tangible interfaces in terms of 'degree of coherence' – i.e. whether the physical and the digital object map onto one another physically and conceptually. They put particular focus on how digital and physical objects can be computationally coupled. Fishkin (2004) provides a two-dimensional classification. He suggests two axes, metaphor and embodiment, as particularly useful when describing and analyzing tangible interfaces. Fishkin's embodiment dimension represents the extent to which the input focus is tied to the output focus in a TUI application. The second axis in Fishkin's framework is metaphor. In terms of tangible interfaces this means the strength of analogy between the interface and similar actions in the real world. TUIs may be a solution for providing innovative ways for children to play and learn, through novel forms of interacting and discovering and the capacity to bring the playfulness back into learning (Price et al., 2003). Marshall (2007), on the other hand, identifies six perspectives with respect to TUIs: possible learning benefits, typical learning domains, exploratory and expressive types of activity, integration of representations, concreteness and sensory directness, and effects of physicality on learning. Marshall also identifies a number of possible learning benefits of interacting with TUIs that have been suggested but not tested: learning benefits of physicality, collaboration, accessibility, novelty of links and playful learning. Antle (2007), with her CTI framework recommends body-based understanding of concepts and spatial schema for more abstract concepts. She also considers that, children's understandings of cause and effect relations can offer learning opportunities. Price (2008) interprets tangibles as representational artifacts and present taxonomy for conceptualizing tangible learning environments with respect to issues of external representation. Antle and Wise (2013), through her Tangible Learning Design Framework argue that TUIs present unique opportunities to support learning interactions. According to the above framework we can reduce extraneous cognitive load by making mappings between the form and behavior of physical and digital objects. We can also use conceptual metaphors in order to structure interaction mappings for supporting learning of abstract concepts. Manches et al. (2009) argue that digital manipulatives (digitally augmented physical objects) present opportunities to transform young children's learning by bridging their pre-school concrete experiences with more formal, abstract, concepts.

4. Material and Methods

The research question around which the study has been designed was the role of conceptual metaphors in improving children's ability for solving addition problems. The study used a between subjects' design with conceptual metaphor (No/Yes) as the independent variable. All children solved ten addition problems (equalize type). The dependent variable was the number of correct solutions.

4.1. Participants

The study took place, halfway through the school term in two primary and two nursery suburban state schools in the Greek city of Larissa. Participants were aged 5 to 7 years of age (24 girls and 22 boys). Children were chosen from Reception and Year 1 class. All children had Greek as their first language and none had any special needs. Ethics approval forms explaining our research goals and additional documents were submitted prior to the study, to the Institute of Educational Policy (IEP) Greece. After obtaining approval from the IEP we also obtained consent of interest from the participants' legal guardians. Parents of children signed and returned the consent of interest form about permitting their child to take part in our research study.

4.2. Materials and Procedure

Sessions took place in classrooms in a familiar classroom adjoining one of the classrooms. There was one child per each session. Noise levels and distractions were kept low. Sessions lasted between six and twelve minutes. The researcher, who was unfamiliar to the children, spent a few minutes conversing with each child to put him/her at ease before sessions began. The researcher began by welcoming the child and thanking him/her for his/her participation. It was then explained in general terms that the aim of the study was to find out what helps children learn about numbers. Questions were presented in the same order to all children.

The researcher narrated the problem story that involved a child, his/her two animals (a frog and a turtle) and a couple of sweets. According to the story, the frog picked one sweet while the turtle picked three sweets. The child was asked what he/she would have to do so that the frog could have as many sweets as the turtle. The researcher used an image of the animals and cubes (2.3 cm X 2.3 cm X

2.3 cm) to represent the sweets to help the children visualize them. The researcher demonstrated how the cubes should be placed on the table. For the study we developed an interactive tabletop for object tracking, based on visually identifying the objects. A printed pattern is attached to objects in order to help the system to visually recognize them through a digital camera. The table, which has been custom built from cardboard is 46 cm long, 35 cm wide and 40 cm tall, and has been designed to be transportable. The height of the table can be adjusted according to the height of the child. A camera (Sony PS3eye) with wide-angle lenses was placed underneath the table in an unobtrusive location. In addition, one light source was used to illuminate it. On the top side of the cardboard box, a perspex board (51 cm long by 50.5 cm wide) was used with some ordinary tracing paper as shown in Figure 1.

Figure 1: Tabletop with physical manipulatives

This material is completely transparent for objects. In order to avoid direct reflections of the light source, the lower side of the surface has a matte finish, while maintaining the overall transparency. Fiducial markers attached onto red cubes, used to enable place tracking of objects as shown in Figure 2.

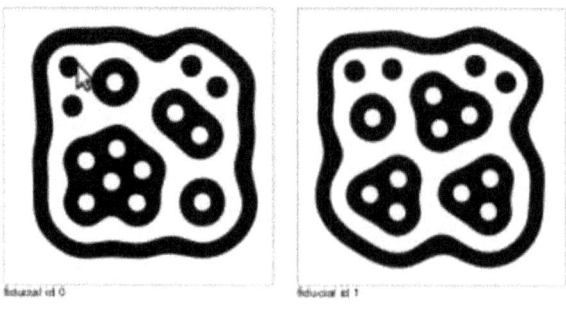

Figure 2: Fiducial markers

We used reacTIVision which is an open source, cross-platform computer vision framework for the fast and robust tracking of fiducial markers (Kaltenbrunner and Bencina 2007). Optimized image recognition algorithms translate the fiducial to a digital identification on the interactive application. Also, image algorithms track the position of multiple fiducials placed on the tabletop surface. ReacTIVision, is a standalone application, which visually recognize fiducials, analyzing the image captured by a camera connected to the system and sends. The TUIO protocol was adopted for encoding the state of tangible objects from our interactive tabletop. We developed our application by using Processing as programming language. Processing is a programming language based on a development environment and an online community. TUIO is an open framework that defines a common protocol and API for tangible multitouch surfaces. The TUIO protocol allows the transmission of an abstract description of interactive surfaces, including touch events and tangible object states. Processing TUIO Client API is part of reacTIVision, an open source fiducial tracking and multi-touch framework based on computer vision. Our application communicated with TUIO protocol via Processing TUIO Client API which serves as the basis for the development of tangible user interface applications. Perspex board which was on the top side of the cardboard box was with black paper except for two equal areas (21 cm X 13.5 cm) either side of the centerline. In front of the left area an image of a frog (13 cm X 10 cm) and in front of the right area an image of a turtle (13 cm X 10 cm) were placed in order to better support children's understanding. A cube placed in the left area means that this sweet belongs to frog. If a cube is placed by a child in the right area means that a sweet is given to turtle. When children place one red cube on the left area, one red square is appeared on the left pan on the monitor as shown in Figure 3. The pan tilts to the side that has more squares. When both sides have the same number of squares the pan balances.

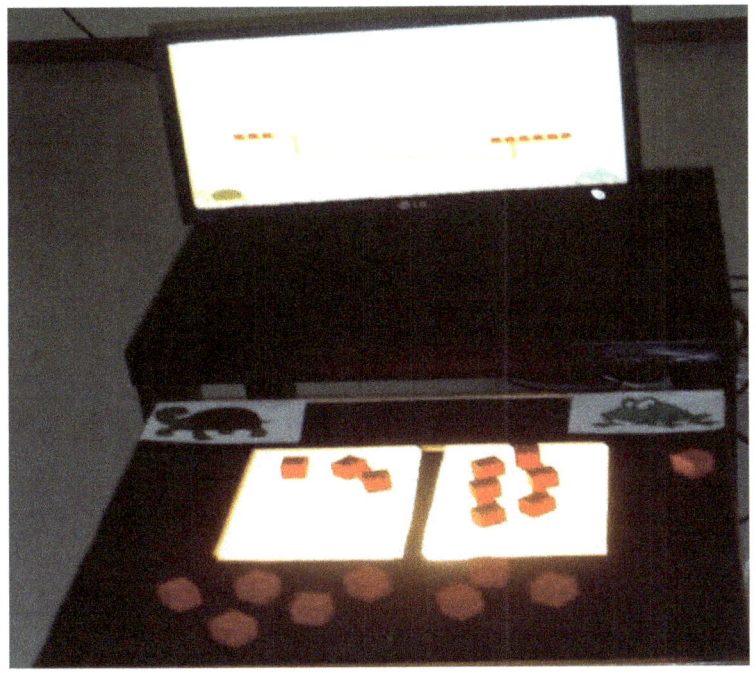

Figure 3: Tangible user interface system

5. Results

The experiment involved 46 subjects who were between five and seven years old from Reception and Year 1 (24 girls and 22 boys). Data on age and gender were collected. Data processing and statistical analysis were performed using the SPSS Statistics Software package (version 17). A Mann-Whitney test revealed significant differences between correct scores in the Physical (Mean=8.09) and Tangible conditions (Mean=9.63).

Table 1: Mean correct scores for Physical and Tangible conditions

Condition	Mean	Std. Deviation	Std. Error of Mean	N
Physical	8.09 b	1.444	.308	22
Tangible	9.63 a	.576	.118	24
Mann-Whitney	P<0.001			
Total	8.89	1.320	.195	46

Mann-Whitney tests showed that although Year 1 children identified more correct solutions than Reception children, the difference was not significant for either the Physical or Tangible conditions.

Table 2: Mean correct scores for Physical and Tangible conditions for children in Reception and Year 1

Condition	Age	Mean	Std. Deviation	Std. Error of Mean	n
Physical	Reception	7.57 a	1.718	.649	7
	Year 1	8.33 a	1.291	.333	15
	Mann-Whitney	P=0.395			
Tangible	Reception	9.67 a	.707	.236	9
	Year 1	9.60 a	.507	.131	15
	Mann-Whitney	P=0.662			
Total	Reception	8.75	1.612	.403	16
	Year 1	8.97	1.159	.212	30

6. Discussion

As shown in Table 1, children that used TUI were able to deliver more correct answers (9.63) on average, compared to children that used only tangible interfaces (on average 8.09 correct solutions). As Table 2 indicates, the children's performance is not affected by age. No difference of statistical significance was observed in the performance of Reception class children compared to that of Year 1 children. Previous research shows benefits for learning through physical interactions in virtual spaces embed children as elements within the systems they are attempting to learn (Birchfield et al., 2008; Lindgren & Moshellto, 2011; Price & Rogers, 2004). This study aimed to identify important parameters for the design of innovative computer-based learning environments used for mathematical education and in particular addition problems. More specifically we studied the effect that tangible interfaces and concept metaphors have on learning. We compared the performance of students when only tangible interfaces were used, to that when digital representations of the interfaces were used. Our research hypothesis was that the external representation and the conceptual metaphor of the sea saw would help students solve correctly more addition problems. Results indicate that tangible interfaces that utilise interactions with external digital metaphors can help students to improve their performance in mathematical problem solving. One of the limitations of the study involves the customised nature of the interactive tabletop that makes it difficult to reproduce in different experimental settings. Another limiting factor is that each child carried out the experiment in isolation, i.e., without physical/social interaction with other children from his/her class. It would be interesting to extend our study in conditions where the experiment involved groups of interacting children.

If children are prematurely introduced into abstract symbolic systems, they will flounder (Mattthews & Scow, 2007). Children need to experience a wide range of interfaces utilizing multiple sensorial dimensions such as tangible interfaces in addition to pencils for sketching, in order to support cognitive development in the early stage of their life (Kim & Cho, 2014).

7. Conclusions

Many researchers have suggested that tangible user interfaces (TUIs) might have significant impacts on learning for a variety of interrelated reasons. They promote exploratory and expressive activities, provide opportunities for reflection, offer learning of abstract concepts through concrete representations and allow natural form of interaction. However, little empirical work exists that provides evidence for enhanced learning in mathematics, specifically for solving addition problems by young children (5-7 years). Our research has shown that children can identify more correct solutions using conceptual metaphors compared to when no metaphors are employed. In order to evaluate the potential for tangibl0e technologies to support children's numeric development, it is important to first identify the possible advantages, as well as limitations, of using conceptual metaphors in this domain.

REFERENCES

Antle, A. (2007). The CTI Framework: Informing the Design of Tangible Systems for Children. In B. Ullmer, & A. Schmidt (Eds.), *Proceedings of the 1st international Conference on Tangible and Embedded interaction-TEI '07* (pp. 195–202). ACM Press.

Antle, A., & Wise, A. (2013). Getting down to Details: Using Learning Theory to Inform Tangibles Research and Design for Children. *Interacting with Computers 25*(1), 1–20. https://doi.org/10.1093/iwc/iws007

Birchfield, D., Thornburg, H., Megowan-Romanowicz, C., Hatton, S., Mechtley, B., Dolgov, I., & Burleson, W. (2008). Embodiment, Multimodality, and Composition: Convergent Themes Across HCI and Education for Mixed-Reality Learning Environments. *Journal of Advances in Human-Computer Interaction 2008*, 1-19. https://doi.org/10.1155/2008/874563

Dourish, P. (2001). *Where the Action Is. The Foundations of Embodied Interaction*. MIT Press.

Fishkin, P. K. (2004). A Taxonomy for and Analysis of Tangible Interfaces. *Personal and Ubiquitous Computing, 8*(5), 347–358. https://doi.org/10.1007/s00779-004-0297-4

Kaltenbrunner, M., & Bencina, R. (2007). ReacTIVision: A Computer-vision Framework for Table based Tangible Interaction. In B. Ullmer, & A. Schmidt (Eds.), *Proceedings of the 1st international Conference on Tangible and Embedded interaction-TEI '07* (pp. 69–74). ACM Press.

Kim, M. J. & Cho, E. M. (2014). Studying Children's Tactile Problem-solving in a Digital Environment. *Thinking Skills and Creativity, 12,* 1–13. https://doi.org/10.1016/j.tsc.2013.11.001

Koleva, B., Benford, S., Ng, K. H., & Rodden, T. (2003). A Framework for Tangible User Interfaces. In L. Chittaro (Ed.), *Proceedings of the Fifth International Conference on Human*

Computer Interaction with Mobile Devices and Services-Mobile HCI 2003 (pp. 46–50). Springer Verlag.

Lindgren, R., & Moshellto, M. J. (2011). Supporting Children's Learning with Body based Metaphors in a Mixed Reality Environment. In T. Moher, C. Quintana, & S. Price (Eds.), *Proceedings of the 10th International Conference on Interaction Design and Children IDC '11* (pp. 177–180). Association for Computing Machinery

Manches, A., O'Malley, C., & Benford, S. (2009). Physical Manipulation: Evaluating the Potential for Tangible Designs. In N. Villar, S. Izadi, M. Fraser, & S. Benford (Eds.), *Proceedings of the 3rd International Conference on Tangible and Embedded Interaction* (pp. 77–84). ACM.

Marshall, P. (2007). Do Tangible Interfaces Enhance Learning? In B. Ullmer, & A. Schmidt (Eds.), *Proceedings of the 1st international Conference on Tangible and Embedded interaction-TEI '07* (pp. 163–170). ACM Press.

Matthews, J., & Seow, P. (2007). Electronic Paint: Understanding Children's Representation through their Interactions with Digital Paint. *Journal of Art Design, 26*(3), 251–263. https://doi.org/10.1111/j.1476-8070.2007.00536.x

Price, S. (2008). A Representation Approach to Conceptualizing Tangible Learning Environments. In A. Schmidt, H. Gellersen, E. van den Hoven, A. Mazalek, P. Holleis, N. Villar (Eds,), *Proceedings of the 2nd International Conference on Tangible and Embedded interaction-TEI '08 (pp.* 151–158). ACM Press.

Price, S., & Rogers, Y. (2004). Let's Get Physical: The Learning Benefits of Interacting in Digitally Augmented Physical Spaces. *Computers & Education, 43*(1–2), 137–151. https://doi.org/10.1016/j.compedu.2003.12.009

Price, S., Rogers, Y., Scaife, M., Stanton, D., & Neale, H. (2003). Using 'tangibles' to Promote Novel Forms of Playful Learning. *Interacting with Computers, 15*(2), 169–185. https://doi.org/10.1016/S0953-5438(03)00006-7

Ullmer, B., & Ishii, H. (2001). Emerging Frameworks for Tangible User Interfaces. In J. M. Carroll (Ed.) *Human-Computer Interaction in the New Millenium* (pp.579–601). Addison-Wesley.

Chapter 10

Utilizing Sphero for A Speed Related STEM Activity in Kindergarten

Michalis Ioannou[1] & Tharrenos Bratitsis[1]

[1]*University of Western Macedonia, mioannou@uowm.gr, bratitsis@uowm.gr*

Abstract: *STEM education is being gradually adopted by all levels of Education. Kindergarten lately attracts more attention from the STEM education policy makers because it is believed that children who develop an interest in STEM at a young age are more likely to excel in the future and avoid stereotypes or other obstacles when entering STEM fields in later years. In this paper the design of a teaching activity for approaching the notion of speed in Kindergarten, utilizing the Sphero SPRK robot, is described. The proposed activity is part of a postgraduate dissertation, still in progress.*

Keywords: *STEM, Kindergarten, Speed, Sphero*

1. Introduction

STEM was firstly introduced in the 1990's by the National Science Foundation and has been used since, as a generic label for any action, policy, program or practice that involves one or more of the STEM disciplines: Science, Technology, Engineering, & Mathematics (Bybee, 2010). STEM education can be defined as an integrative approach to curriculum and instruction, content and skills, approaching all the STEM areas as one, without any boundaries between them (Maryland State Department of Education, 2003; Morrisson & Bartlett, 2009). Through STEM education, students can develop 21st Century skills like adaptability, problem solving, complex communication and system thinking (National Research Council [NRC], 2010). Generally, STEM education seems to have some benefits, as students become better problem solvers, innovators, logical thinkers, inventors, and technologically literate (Morrisson 2006). Also, students become STEM literate through this kind of educational approaches. STEM literacy includes the conceptual understanding and procedural skills and abilities for individuals to address not only personal, but also social, and global issues (Bybee, 2010). Recently, a new tendency is evident, suggesting the exploitation of artifacts in STEM education, for fostering creativity and innovation among students through a more attractive way of STEM education. Thus, Arts are proposed as an additional constituent, leading to the generation of the STEAM term (Science, Technology, Engineering, Arts & Mathematics; Stemtosteam, n.d.).

It seems that STEM programs can be implemented in Kindergarten. Children from a very early age formulate theories and ideas for just about everything, and these ideas play a significant role in the learning experience (Jos, 1985). Even before they become involved in any educational system, they form ideas for a variety of physics' phenomena around them, thus constructing definitions about them. Through community, and social interactions children look up for definitions of the world around them (Driver, 2000). Also, according to the recent and growing recognition of the role of stimulation in early brain development, it seems that preschool training programs and Kindergarten in particular, provide a significant place to start focusing on STEM education in order to obtain positive fruitful results in the future (Torres-Crespo et al., 2014).

The proposed activity is part of a postgraduate dissertation in progress. In this paper the design of STEM related activities about the notion of speed in Kindergarten, is described. More specifically, activities with the robot Sphero SPRK, are suggested in order to introduce the notion of speed in Kindergarten. The present paper is structured as follows: firstly, the theoretical framework is presented. Then, the proposed activities are described, followed by a concluding discussion.

2. Theoretical Framework

Speed is defined as the distance covered per unit of time (speed = distance/time). It was first measured by Galileo, considering the distance covered and the time needed to achieve that. Velocity combines both the ideas of speed and direction of motion. In general, speed is a description of how fast an object moves and velocity is how fast it moves and in which direction (Hewitt, 2015). In the New Curriculum for Kindergarten in Greece (New Curriculum for Kindergarten in Greece, 2011), speed can be found in Physics Science section, specifically in the "Notions and phenomena from the natural world" part, under the topic "Simple physics phenomena about: the motion of objects". Also in the Interdisciplinary Frame Curriculum for Kindergarten in Greece (Interdisciplinary Frame Curriculum for Kindergarten in Greece [I.F.C.], 2003) speed can be found in section "Child and Mathematics", specifically in the "children should be able to compare the speed of objects' motion compared to time when the distance is same" part and in the "Natural Environment and interaction" topic. In the latter case, the pursued ability to be cultivated is "to perceive the motion and its principles".

Inquiry, as a term, is used throughout the science education literature to describe goals for science learners as well as approaches for science teaching (NRC, 2000). Students who use inquiry to learn science are engaged in many activities and thinking processes which are similar to those scientists do (NRC, 1996). Classroom inquiry can occur at all grade levels of education (NRC, 2000). "Inquiry is a multifaceted activity that involves making observations; posing questions; examining books and other sources of information to see what is already known; planning investigations; reviewing what is already known in light of experimental evidence; using tools to gather, analyze, and interpret data; proposing answers, explanations, and predictions; and communicating the results. Inquiry requires identification of assumptions, use of critical and logical thinking and consideration of alternative explanations (NRC, 1996).

Some practices that can take part in K-12 science classrooms are: a) Asking questions (for science) and defining problems (for engineering), b) developing and using models, c) planning and carrying out investigations, d) analyzing and interpreting data, e) using mathematics and computational thinking, f) constructing explanations (for science) and designing solutions (for engineering), g) engaging in argument from evidence, and h) obtaining, evaluating and communicating information (NRC, 2012).

Scientific investigation, inquiry and engineering design are closely related activities that can sometimes be mutually reinforcing in many curricula. Scientific inquiry and engineering design are often compared to each other, because of the problem-solving approach that they both use. It seems that certain science concepts including the use of scientific inquiry methods can support engineering design activities (NRC, 2009). The engineering design process offers a context which can support educators when teaching inquiry and scientific reasoning. The engineering design process is compared with the scientific inquiry process, as shown in Table 1 (Rockland et al., 2010):

Table 1: Comparison of the Engineering Design Process and the Scientific Inquiry/investigation Process

Engineering Design	Scientific Inquiry
1. Identify the need or problem	1. Formulate the problem
2. Research the need of problem	2. Information gathering
3. Develop possible solutions	3. Make hypotheses
4. Select the best possible solution	4. Plan the solution
5. Construct a prototype	5. Test solutions (perform experiments)
6. Test and evaluate the solution	6. Interpret data. Draw conclusions
7. Communicate the solution	7. Presentation of results
8. Redesign	8. Develop new hypotheses

EiE (Engineering is Elementary; EiE, 2016) is an inquiry-based STEM curriculum that teaches students a variety of skills including problem solving and thinking skills and it is built around and engineering design process. It is developed by the Museum of Science in Boston and proposes a simple five-step Engineering Design Process to guide students, especially children at a young age, through any engineering design challenges. Figure 1 presents the Engineering Design Process which is a flexible cycle without a start or an end point. Anyone is able to begin with any step from: a) ASK (a problem is faced), b) IMAGINE (brainstorming solutions), c) PLAN (diagram and materials), d) CREATE (follow the plan or test it), and e) IMPROVE (changes and modifications.

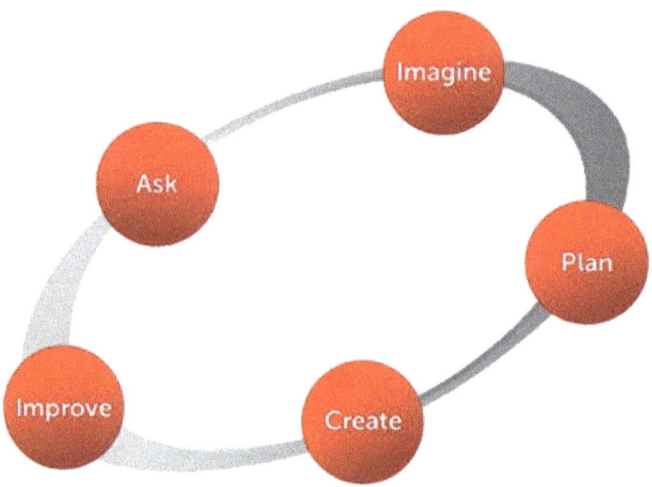

According to Havice (2009), implementing teaching strategies, like problem-based learning within a STEM curriculum, may enhance students' desire to understand the world around them and engage them in classroom instruction. Skills like problem solving, literacy, creativity, and motivation are positively influenced when children access technology in their learning environments. Also, using technology as an instructional tool enhances children's learning and educational outcomes (New Curriculum for Kindergarten in Greece, 2011). According to Resnick (2007) considering technology not only as a supporting tool, but mainly as a creative tool for current teaching methods offers children the opportunity to evolve as creative thinkers and explorers who plan, analyze, and share their work collaboratively.

STEM education is described as a center of integrated disciplines, as an interdisciplinary bridging among discrete disciplines or as an entity (Morrisson, 2006). Sanders states that the "notion of integrative STEM education includes approaches that explore teaching and learning between/among any two or more of the STEM subject areas, and/or between a STEM subject and one or more other school subject." (Sanders, 2009).

The early education system needs to shift from fearing STEM to playing with it. Teachers in the preschool or the early education system are probably already applying parts of STEM in everyday basis without realizing it. It is the educators' responsibility to work with children during these early years by using educational appropriate strategy and playful approaches to foster STEM skills (Torres-Crespo et al., 2014).

Children must develop an interest in STEM at a young age to excel at them when they are older (Hunter, 2015). Researchers and policy-makers have pushed for an increased focus on STEM in early childhood. Research, also, has highlighted the importance of exposing young children to STEM in an early age in order to avoid stereotypes or other obstacles to enter STEM fields in later years (Elkin et al., 2014; Jos, 1985).

Finally, robotics as a tool can help make abstract ideas more concrete. Children can directly view the impact of their programming commands on the robot's action. Children who work with robots can move around the room, work on the floor or a table, and act out with their own bodies, before programming the robot. Robotics becomes a new early childhood useful educational mean of the 21st Century (Sullivan et al., 2013).

3. Goal Setting Activities

The proposed activities concern teaching the notion of speed in Kindergarten by utilizing the programmable robot "Sphero SPRK" (http://www.sphero.com/) (Figure 2). Sphero is a remote-controlled robotic ball, designed by Orbotix. It is capable of moving around in any direction and in various speeds. In this paper the Lightning Lab application which runs on a tablet is proposed to be used in order to program Sphero's movement. Specifically, the introduction of Sphero in the Kindergarten classroom is suggested, following the implementation of two additional activities in which children take action with their bodies. The activities have been designed within the context of a postgraduate thesis, still in progress, and will be implemented in early-December, 2016.

Figure 1: Sphero SPRK

In the first activity, children will realize that the faster competitor in a fixed-distance race is the one who needs the shortest time to reach the finish. Through this activity children are required to solve a problem that animals living in a forest face. Specifically, they need to know the winner of a race among them with a designated, common start and termination point. Thus, children should be able to search for an answer to questions like "Who is the winner if the distance is the same?", "Who finishes first?". In order to achieve that, children will be required to organize their own race. The researchers will guide the activity introducing two important tools, the chronometer and a measuring instrument for distance which will be decided together with the children after an in-class discussion (it can be a ribbon, a stick or anything they choose). The chronometer is designed in the Scratch environment (http://scratch.mit.edu/) and uses both numbers written on a PC screen and verbal representation. Additionally, hand-claps will be used to count time with the help of chronometer. Children will be requested to run individually or in dyads, while the others count their time using hand-claps and all together will collect and write down the data related to the time each child needs to reach the finishing point, on a board. At the end of the activity, children will be requested to compare their collected data and decide which one of the competitors is faster, but also to justify their answer.

In the second activity, it is expected that children will realize that the faster is the one who covers the longer distance in the same period of time, as the others. More specific, children will be confronted with another problem that the animals face. "What if we race without having a distance limit, a specific route? Who wins then?". Children need to design a different race in this time case. They will be challenged to cover (run/walk) a non-limited distance in a certain amount of time. They run individually and/or in groups. When the time ends, they will be asked to place custom-made sing with their name on the floor, next to their feet and measure the distance with the measuring instrument they will have already agreed upon, at the beginning of the intervention. Individually they will be asked to collect and note their data (distance). At the end of the activity, the children will be requested to compare their data with each other's and decide, also justifying their conclusion, who is faster.

Children in both activities will have to work with the notions of distance and time, while trying to find out and justify "who is faster or slower". For reaching an understanding and formulating their answers, children will be required to organize and compete in their own races as aforementioned. Through the last activity it is expected that a strong representation for speed will be acquired by the children, who will connect the terms "faster" and "slower", which they are already familiar with at this age, with the notion of speed. In this activity, an attempt will be made to introduce the notion of speed related to "how fast someone or something moves", with Sphero SPRK. Children will be able to program Sphero's speed through a tablet, using an application called "Lightning Lab" (Figure 3a) and see in action what happens if Spheros' speed is altered. Moreover, the researchers will introduce the notion of speed in terms of "faster" and "slower". The researchers will highlight that "the notion of speed, according to scientists, is how fast an object moves. If someone or something is faster than someone or something else, this means that his/its speed will be higher". In order to examine this statement, children will be introduced to Sphero, "a tool that scientists use to examine speed". A scientific experiment is planned to take place in the Kindergarten, using Sphero, utilizing the new knowledge of speed from the previous activities. Sphero will be controlled by children, through the Lighting Lab application, using a scale made by the researchers which utilizes different animals in order to make it easier for children to understand speed through metaphorical representation (the animals can be easily ordered according to how fast they are, Figure 3b). Children will control the speed of Sphero recognizing the five-scaled representation using 1 for the lowest and 5 for the highest speed. The researchers use animals as well as numbers in an appropriate redesigned environment (using stickers on the screen of the tablet) in order to make it easier for children to understand speed grading. Children will organize with the support of the researchers two experiments with Sphero: a) Sphero covers same distances with different speed, and b) Sphero covers different distances in a same amount of time. Through this activity children will be able to recognize: a) when the speed of Sphero is higher the designated distance is covered faster (experiment a), and b) when the speed of Sphero is higher it covers longer distances in the same amount of time (experiment b).

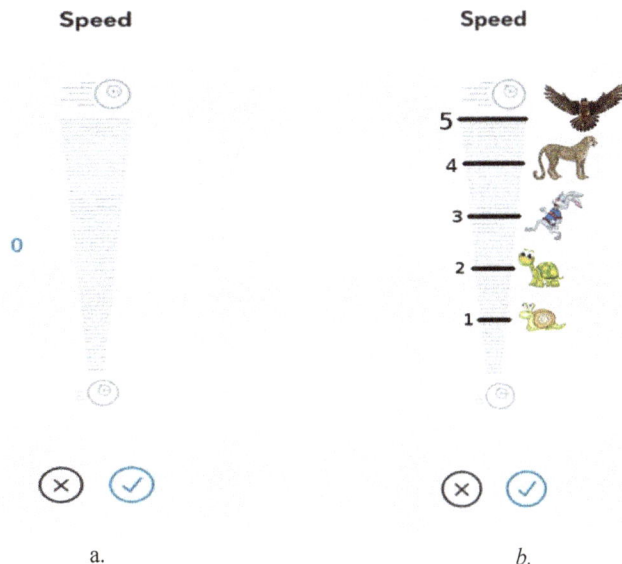

Figure 2: Lightning Lab – Speed

4. Discussion

With the proposed activities, it is supposed that children will approach an understanding of the notion of speed. They should be able to not only define which of two objects/beings is faster, but also to predict which will be faster if we know the speed of the two objects. In addition children should reach an understanding of how fast an object is, using its speed as a justification approach.

The core idea of the proposed intervention is that an empirical experiment will take place in the third activity, involving the Sphero. By setting its speed, they will be able to observe how fast it is. Through the first two activities, the children will work with notions they are already familiar with at this age; time and distance. They will be facilitated to connect them with the terms "faster" or "slower", which they also are familiar with. In an attempt to reinforce their perception, the race setting has been designed, also supported by questions like "who will finish first" and "who covered the longer distance".

Through the third, experimental activity, the children will be required to apply their perceived knowledge from the first two activities in order to attempt to understand the notion of speed. The research hypothesis is that they will be able to understand that speed is connected to how fast someone of something is, that speed is connected to distance covered and time required to do so. At this stage it is not easy to predict how well they will perceive the notion of speed, but if the connection among the core constituents can be made (speed, time, distance), the intervention will be considered as successful. The children are not ready to understand the mathematical relation between those elements, so an understanding that there is a relation and an empirical recognition of "who/what is faster than the other" will be considered as adequate for this age, since speed is not being discussed at such young ages.

By evaluating the study implementation, the researchers plan on being able to determine the children's readiness to understand the designated notions and identify the element which facilitates this understanding more (animal representation, race setting, facilitating questions, etc). The latter will be evaluated through semi-structured interviews with the children and observations. The results will determine future research paths for the same or similar notions in the science field.

REFERENCES

Bybee, R. (2010). Advancing STEM Education: A 2020 vision. *Technology and Engineering Teacher, 70*(1), 30–35.
Driver, R. (2000). *Children's Ideas in Science.* Open University Press.
EiE (2016). *Engineering is Elementary.* Rerieved October 10, 2016, from http://www.eie.org.
Elkin, M, Sullivan A., & Bers, M. U. (2014). Implementing a Robotics Curriculum in an Early Childhood Montessori Classroom. *Journal of Information Technology Education: Innovations in Practice, 13*, 153–169.

Havice, W. (2009). The power and promise of a STEM education: Thriving in a complex technological world. In ITEEA (Ed.), *The Overlooked STEM Imperatives: Technology and Engineering* (pp. 10–17). ITEEA.

Hewitt, P. G. (2015). *Conceptual Physics* (12th ed.). Pearson Education.

Hunter, J. (2015). *STEM education: Kindergarten is where it should begin.* Retrieved October 5, 2016, from http://www.smh.com.au/comment/stem-education-kindergarten-is-where-it-should-begin-20150816- gj00p5.html

Interdisciplinary Frame Curriculum for Kindergarten in Greece (2003). *I.F.C. for Kindergarten.* Hellenic Ministry of Education, Lifelong Learning and Religious Affairs.

Jos, E. (1985). The right question at the right time. In Harlan, W. (Ed.), *Primary Science: Taking the Plunge.* Heinemann.

Markert, L. R. (1996). Gender related to success in science and technology. *The Journal of Technology Studies, 22*(2), 21–29.

Maryland State Department of Education. 2003. *Maryland STEM.*

Morrison, J. (2006). *TIES STEM education monograph series: Attributes of STEM education.* TIES.

Morrison, J., & Bartlett, R. V. (2009). STEM as a curriculum: An experimental approach. *Education Week, 23,* 28–31.

National Research Council (1996). *National Science Education Standards.* National Academies Press.

National Research Council (2000). *Inquiry and the National Science Education Standards: A Guide for Teaching and Learning.* National Academies Press.

National Research Council (2009). *Engineering in K-12 Education, Understanding the status and improving the prospects.* National Academies Press.

National Research Council (2010). *Exploring the intersection of science education and 21st century skills: A workshop summary.* National Academies Press.

National Research Council (2012). *A framework for K-12 Science Education: Practices, Crosscutting Concepts, and Core Ideas.* National Academies Press.

New Curriculum for Kindergarten in Greece (2011). *Curriculum for Kindergarten 2011.* Act "New School (21st Century School) New Curriculum, Priority Axes 1,2,3 Horizontal Action" with MIS code 295450, Subproject 1: Development of Curriculum for Primary and Secondary Education and guidance for the teacher 'Tools for teaching Approaches' (2011). Pedagogical Institute.

Ohlson, T., Monroe-Ossi H., Fountain, C., McLemore, B., Carlson, D., & Wehry S. (2016, June 26–29). *Exploring Programming and Robotics in Early Childhood Classrooms* [Paper presentation]. 2016 International Society for Technology in Education: (ISTE) Annual Conference and Exposition, Denver, CO, USA. https://conference.iste.org/uploads/ISTE2016/HANDOUTS/KEY_100471886/OhlsonISTE 2016.pdf

Resnick, M. (2007). Sowing the seeds for a more creative society. *Learning & Leading with Technology, 35*(4), 18–22.

Rockland, R., Bloom, D. S., Carpinelli, J., Burr-Alexander, L., Hirsch, L. S. & Kimmel, H. (2010). Advancing the "E" in K-12 STEM Education. *The Journal of Technology Studies, 36*(1), 53–64.

Sanders, M. (2009). STEM, STEM education, STEMmania. *The Technology Teacher 68*(4), 20–26.

Stemtosteam (n.d.). *What is STEAM?* Retrieved October 10, 2016, from www.stemtosteam.org.

Sullivan, A., Kazakoff, E. R., & Umashi, B. M. (2013). The wheels on the bot go round and round: Robotics curriculum in pre-kindergarten. *Journal of Information Technology Education: Innovations in Practice, 12,* 203–219.

Torres-Crespo, M. N., Kraatz, E., & Pallarsch, L. (2014). From fearing STEM to playing with it: The natural integration of STEM into the preschool classroom. *SRATE Journal, 23*(2), 8–16.

Chapter 11

STEM+ARTS=STEAM Skills: Innovation Management and Scratch Programming for Year 4 Students

Niki Lambropoulos[1] & Ioannis Dimakos[1]

[1]*Human Creativity Unit, Laboratory for the Cognitive Analysis of Language, Learning, and Dyslexia, Department of Primary Education, University of Patras, Greece*

Abstract: *Nowadays, world economic success and well-being require new industries to appear. Technology, Engineering, and Mathematics (STEM) provide tangible solutions towards this direction with STEM introduced in cross-curriculum activities. However, when working in Primary Education, STEM is not enough; Arts and Design (STEAM) can introduce fun, surprise, curiosity, teamwork, co-creativity and innovation. STEAM intervention was conducted with Year 4 students at the 6th Primary School of Patras for one year. Initiated within the Flexible Zone hours, STEAM skills development was anchored in Bruner's proposition: anything can be taught at any age if delivered correctly. STEAM advanced concepts were taught for Innovation Management and Programming with Scratch using metaphors and practical examples. Art was approached via Design Thinking within the Zone of Proximal Flow. The creative classroom activities were orchestrated based on Computer Supported Collaborative Creativity and Learning (CSCC/L) The 10 years old students produced innovations and a Scratch game by having fun, innovating and learning together.*

Keywords: *STEAM, Design Thinking, Zone of Proximal Flow, Scratch, Innovation Management*

1. Design Thinking + Hybrid Synergy

Teachers play a significant role stimulating children's curiosity, imagination and willingness to experiments, helping them develop the transversal competences required for creativity and innovation, such as critical thinking, problem-solving and initiative-taking. Following Piaget (1988), the principle goal of education in the schools should be creating men and women who are capable of doing new things, not simply repeating what other generations have done; men and women who are creative, inventive and discoverers, who can be critical and verify, and not accept, everything they are offered. Be one with the universe, and so very identical to it that he does not even feel the need for two terms. The purpose of education is to facilitate thinking and problem solving skills so to be transferred to a range of situations, also developing symbolic thinking by discovery learning; students construct their own knowledge for themselves (constructivist approach; Bruner, 1960, 1961). According to Csikszentmihalyi (1990, 1996), the secret to happiness is to be in the creative flow state. Hence, there can be an interdisciplinary field where students can be creative and happy achieving the optimal experience in education.

This field is STEM (Science, Technology, Engineering, and Mathematics); but STEM is not enough. It is also essential to support young students' group experience, fun, creativity, curiosity and innovation. Such elements exist in STEM with Arts and Design via collaborative learning environments. STEAM skills development stands on Bruner's proposition that anything can be taught at any age if delivered correctly. Design thinking is a method for creative action and was adapted for solution and user driven approaches. Design thinking employs creativity techniques for idea generation (Eureka moments) and evaluation.

Anchored in Design Thinking, 26 Year 4 students were introduced to advanced concepts in Innovation Management and Scratch Programming. Time was essential in both activities as innovation management requires scheduling and as the students entered a competition with Scratch with deadline the end of March 2015. Hybrid Synergy creativity technique activated both left and right brain hemispheres (LBH, RBH). LBH is objective, analytical, logical and classical; RBH is more informational, holistic, continuous, patterns, gestalt capturing and subjective. In reality, both L/RBH functionalities make our abilities and competencies move from an individual self-perception to a more unified perception of our world (Kounios & Beeman, 2015). Hybrid Synergy (Lambropoulos et al., 2008) is a creativity technique that synthesized De Bono's thinking hats (2000) and critical thinking levels.

Hybrid Synergy is a 5levels non-linear analytical framework that facilitates and enhances e-learners' metacognitive awareness. Hybrid Synergy levels enhancing critical thinking are: information provision; social inferences; ideas exploration; ideas evaluation; and overviews, summaries and task allocations. These levels are nonlinear and not predetermined. Both parts of our brains can function as a unity much faster providing more accurately perception of reality so to respond to a need or a problem. In this way, cognitive plasticity and flexible learning embraces ambiguity, requiring creativity competencies on-the- spot dealing with rapid changes in information, knowledge, signification and meaning. Furthermore, the students were acquiring new competencies as the projects were developing.

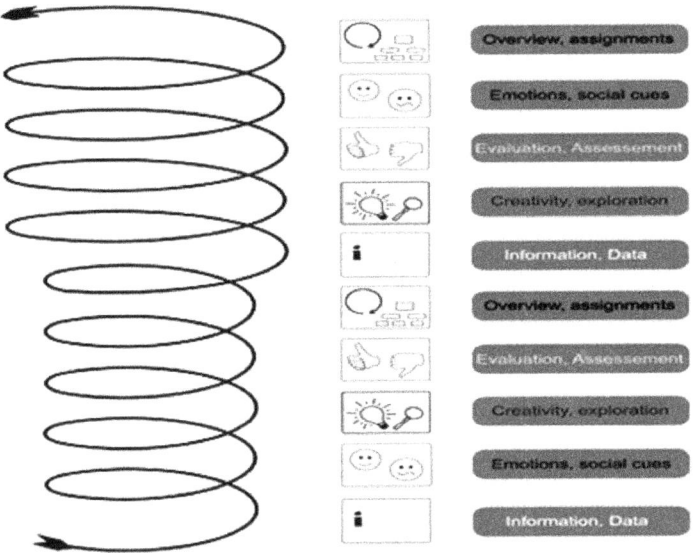

Figure 1: The Hybrid Synergy framework for collaborative creativity

2. Art + Creativity in the Zone of Proximal Flow (ZPF)

Innovation and learning can be collaborative experiences mixing 'older' and more experienced with 'younger' students. Knowledge acquisition and symmetry in groups converge and align asymmetrical interactions between learners and more capable peers, the Zone of Proximal Development (ZPD; Dillenbourg, 1999; Vygotsky, 1978). Creativity is a combination of personal interest and a sense of discordance in the environment, and thus the creative process is a search for interest and novelty by changing the environment to reduce discordance (Martindale, 1990). Flow happens when a person in an activity is fully immersed in a feeling of positive energized focus, full involvement, and success in activity. Ultimate individual or group performance occurs when harnessing the emotions and positively enhanced, channeled, energized, and aligned with the task to promote ultimate learning and performing. Creative flow is a crucial source of internal rewards for humans; it is the self-engagement in activities which require skills just above their current level. Thus, exploratory behavior is explained by intrinsic motivation for reaching situations with a learning challenge. Internal rewards are provided when a situation which was previously not mastered becomes mastered within an optimum amount of time: the internal reward is maximal when the challenge is not too easy but also not too difficult. There are 10 factors to promote the creative flow (Csikszentmihalyi, 1990, 1996):

1. Clear goals where the challenge level and skill level should both be high
2. Concentration and focused attention
3. Loss of feeling
4. Distorted sense of time as in immersion
5. Direct and immediate feedback
6. Balance between ability level and challenge (the activity is neither too easy nor too difficult)
7. Sense of personal control over the situation or activity
8. The activity is intrinsically rewarding, so there is an effortlessness of action
9. Lack of awareness of bodily needs, and
10. Absorption into the activity.

The Zone of Proximal Flow (ZPF; Basawapatna et. al., 2013; Repenning, 2012) is the area where Creative Flow occurs within ZPD. Learners' interest and engagement counteract the anxiety experienced; however, for learners to experience ZPF for an enhanced learning experience, immersion is required. Thus, capturing attention for deep engagement facilitates students' involvement in mental creative flow states. In this case, Design Thinking and ZPF needed an innovative pedagogical approach: Computer Supported Collaborative Creativity and Learning (CSCC/L). Individualized learning based on talent management (Coyle, 2009; Robinson, 2009) with direct observation of the students was also a major part of such study as the +U approach: adding own talents, characteristics and personality traits as complementary skills in each team.

3. Computer Supported Collaborative Creativity + Learning for STEAM

Collaborative learning (by UNESCO n/d) takes place when learners work in groups on the same task simultaneously, thinking together over demands and tackling complexities. Collaboration is here seen as the act of shared creation and/or discovery. Anchored in constructionism, Computer Supported Collaborative Creativity and Learning (CSCC/L; Dillenbourg et al., 1996;

Lambropoulos & Romero, 2015) different group members work on different parts of the project that are brought together, also working all together in all project parts for coherency. CSCC/L provided the learning convergence techniques for orchestrating the educational activities: the macro script related to the Jigsaw Puzzle technique and teaching styles; and the micro script for team communication with Hybrid Synergy dialogues, revolved around generative topics; in this case, Innovation Management and Scratch Programming. The target was to enhance students' STEAM competencies in action. Teaching by generative topics is open without strict use of resources so both the students and the teachers can explore the subject to their interest. Thus, different techniques and methods were tried out for the best practices to appear for the specific context and situation, also expanding it to students' own context such as programming in their own time for their own purposes. Such synthesis of formal and informal learning, and also, onsite and online simultaneously, occurred within the IT Room, donated by Niarchos Foundation.

The combination of real-life projects requires STEAM inter-being, evident in the necessity of a range of disciplines and capabilities. The students were organized in small teams to build specific and common purposes and targets, share responsibility, measure their progress, synthesize their complementary skills, agree on the norms and rules, agree on practices and processes to follow for a successful project to mention a few. Specific theoretical and onsite team building approaches were employed as leaving the students decide the teams on their own failed dramatically. Team projects were built on: (a) team culture, the leverage of expertise of others based on everyone's expertise and advance on others; (b) the shared desire and meaningful ideas to each member; (c) each member's significant contribution; (d) a wide spectrum of expertise; and (e) students' full brain utilization and imagination in a very short period of time. Team creative flow also required the students to resolve any conflicts, work towards team project realization, acknowledge each other's abilities and skills, build relationships of trust and support; otherwise they realized the projects would fail.

4. Generative Topics for STEAM

Innovation Management

Innovation Management is the process of implementing an innovative idea in the real world. Briefly, the stages are: idea generation, idea evaluation and selection, recognize opportunities and challenges, idea realization and evaluation, innovation product for display. More specifically:

> 1. Idea Generation: The students utilized the Hybrid Synergy technique. They worked in groups following the creativity process (Wallas, 1926): (a) Preparation: problem identification, gathering relevant information

and their analysis (b) Incubation: positive treatment of the problem, quantity and originality of ideas, suspension of critical process, cross-fertilization of ideas, stimulate new ideas, enthusiasm and acceptable mental status (c) Illumination: maturation of an idea after a reasonable period of time and through the subconscious ideas as the Eureka moment; (d) Verification: evaluate and verify the best idea for implementation.

2. Idea evaluation and selection: The students, working in teams, had to choose the best idea for implementation based on specific quality criteria they had to settle for the project.

3. Recognizing opportunities and challenges: The ideas the teams agreed were: Trojan Horse (play and clean the floor simultaneously to help your mum), Flying City (an ecological city above the ground), Robotokton (a robot that gathers and kills rubbish in the neighborhood), Recycle Ship (a ships that recycles rubbish in the sea, Lancer Evolution (a robot that waters the plants on the earth or the house), and the Mechanic Animal (Μηχανόζωο, robot that protects all plants and animals from extinction).

4. Idea realization and evaluation: the students finalized their innovations and team-to-team evaluation took place, exchanging ideas and comments for improvement.

5. Innovation product for display and marketing: The students produced posters and flyers about their innovations. They also presented them in front of the school block of the 6th and 19th Primary School students and teachers, and also parents. Lastly, there were presentations into different classrooms and YouTube recordings.

Some students comments were: 'Almost from the beginning of the school year we were dealing with innovations, it was about time to create three-dimensional ones. I was surprised with our fantastic and great innovation! Nothing went wrong because we worked nicely together.' (Eftixia, Flying City). 'I was impressed when we constantly disagreed and agreed again so we finally created a very beautiful building. It is small and simple, but I believe it will teach others to love the nature! (Diana, Flying City). 'I was impressed we built a big robot. I was very happy we managed such a big project and I hope to do again so we have perfect time and also show our abilities!' (Dora, Rompotokton). We created incredible innovations for environmental protection together! With my team, we made a horse that cleans the environment while playing. Our materials were simple, glue, wood, wheels, glass for the eyes, sensors and a sponge for the horse tail. (Nicholas, Trojan Horse).

Scratch Programming

Education today cannot be separated with the ICT based activities. Teaching programming languages to kids influences their way of thinking while personal fulfillment and development is achieved. Code Club volunteers were all students in the Department of Computer Engineering in University of Patras (Anyfanti et. al, 2015). For each lesson the teachers showed all the pupils what they would do next on a whiteboard or playing it at the front of the class. Then they go to work in pairs at the computers, trying to achieve the same task using Scratch. By the end of the educational year, the students could create their own games and debug their own programs utilizing technology safely and respectfully. The Jigsaw Puzzle technique. This game development led the slitting of the class into 4 teams: Coding, Drawing, Sound and Script Creative Writing. Concerning programming as such, the students successfully identified, solved and implemented a number of increasingly complex logical, organizational and even cross-specialty problems. They worked on some core values of programming (both serial and object-oriented), from loops and if clauses, to objects working in parallel and event send and handling. Theory was explained via examples both in-code and using little games. Spatial representation was also played onsite following the classroom axes to explain x and y axes on Scratch. The steps on the floor represented the coding numbers on the interface.

During the first meeting with the Code Club volunteers there were team building techniques such as 'catch the ball' and introducing themselves, what they do and interests. Also, an agreement on norms and behaviors was established. As for a class project in order to participate in a major national competition on Scratch in March 2015, the students agreed to participate. During the first-class project meeting the students were split into two groups selecting a different project game, the circus and the cemetery. Such division created a major competitive climate in the classroom so we, the teachers, decided to change methodology. For this we utilized the Jigsaw Puzzle CSCL technique build upon the game development real expert roles as in the industry.

A new project called The Mystery House combined the previous teams' ideas. Students' complementary and successful participation was ensured, also targeting at the acknowledgment that one groups cannot work without the other. In the initial Game development stages, the decision was upon a haunted house in which the player has to follow clues in hopes of finding the hidden treasure. The students were voluntarily participated in 4 teams: Code-Programming, Visual-Drawing, Music-Audio and Script-Writing. When the basic game outlines, structures and designs were in place, each team created workflow diagrams, furniture sketches, creepy ghostly laughs and riddles. The students were taught advanced concepts' in mathematics taught in High school in Greece. For this we used spatial metaphors and role play games to represent the temporal and special connection in space as with the computer interface. Close to the deadline, they realized that after all this hard work they were not there yet, and some decided to quit. The teachers inspired and encouraged them to continue beyond the breaking point. The deadline day was chaotic, however, the Mystery House was finalized and tested. Finally, year 1 students with their teacher came for the first user testing. Then, the game was submitted for the competition.

The students were really surprised from their game result: "I liked that we all worked together" (Diana). "It was perfect!" (Nicolas). "We made a game!!" (Dionysius). "I drew and others were speaking in microphones and others were using computers and others were writing riddles and we all worked as a team" (Charalambos). "I solved many problems it was very fun!!" (Eftixia). Many students created their own games at home, visited the CodeClub website to download more advanced learning material and collaborated between them for more games.

5. Conclusions + Future Trends

An important factor for the current crisis and future development is the provision of innovative opportunities and prospects to young students who may feel threatened and without sustainable future. Group thinking, fun, creativity and production, can provide a completely different viewpoint for school activities. By incorporating Arts and Design in STEM, a STEAM initiative was introduced to Year 4 (10 years old) Primary School students. Computer Supported Collaborative Creativity and Learning (CSCC/L) was the pedagogical methodology to orchestrate an educational environment so to support and enhance STEAM knowledge, skills and competences in practice. Design Thinking and the Zone of proximal Flow, a combination of the creative flow and ZPD were the macro scripts for small groups CSCC/L convergence for: diverse students' learning styles; teaching and learning for left and right brain functionalities; techniques to enhance imagination and creativity, concentration and attention; digital skills development; close cooperation between the teachers and the students with the teachers; and cooperation with the Parents' Association so the students will continue to practice and develop their skills beyond the initiatives. Taking part in a challenging creative flow enhanced by interactive collective intelligence, the interplay between reasoning and curiosity brought dreams to reality. Human creativity is to dream, to imagine, to feel and create together, to change a youngster's behavior, and thus, his/her own destiny.

Acknowledgments

Our thanks go to University of Patras graduates Ioanna Anyfanti, Konstantinos Vasileiadis, Mira Zeit, and Andriani Vgenopoulou, as well as the CodeClub coordinator Nomiki Koutsoumpari. Special thanks to Year4 students (academic year 2015-2016) for their courage to undertake such great effort taking results beyond any expectations. Also, the teachers from the 6th and 19th Primary Schools and everyone who facilitated and empowered the students; the School Parents' Association and Niarchos Foundation for the IT Lab. Last but not least, we would like to thank the Head Teacher, Dr Marianna Bartzakli, for her help and support.

REFERENCES

6th State Primary School of Patras (2018). http://blogs.sch.gr/6dimpat/
Anyfanti, I., Vasileiadis, K., Zeit, M., Vgenopoulou, A., Mpartzakli, M., & Lambropoulos, N. (2015). Computer Supported Collaborative Learning In Small Teams For Scratch: Programming Skills For Year 4 Students At The 6th Primary School Of Patras, Greece. In A. Szűcs & M. Ildikó (Eds.), Proceedings of the EDEN Open Classroom Conference 2015 (pp. 305–312). European Distance and E-learning Network
Basawapatna, A., Repenning, A., Koh, K. H., & Nickerson, H. (2013). The zones of proximal flow: guiding students through a space of computational thinking skills and challenges. In B. Simon, A. Clear, Q. Cutts (Eds.), *Proceedings of the 9th Annual International ACM Conference on International Computing Education Research ICER '13* (pp. 67–74). ACM.
Bruner, J. (1960). *The Process of Education*. Harvard University Press.
Bruner, J. (1961). The Act of Discovery. *Harvard Educational Review, 31*, 21–32.
Code Club: https://www.codeclub.org.uk/about
Coyle, D. (2009). *The Talent Code: Unlocking the Secret of Skill in Sports, Art, Music, Math, and Just About Anything*. High Bridge Company
Csikszentmihályi, M. (1990). *Flow: The Psychology of Optimal Experience*. Harper & Row.
Csikszentmihalyi, M. (1996). *Creativity: Flow and the Psychology of Discovery and Invention*. Harper Perennial.
De Bono E. (2000). *Six Thinking Hats*. Penguin.
Dillenbourg P. (1999). What do you mean by collaborative learning? In P. Dillenbourg (Ed.), *Collaborative-learning: Cognitive and Computational Approaches* (pp. 1–19). Elsevier.
Dillenbourg, P., Baker, M., Blaye, A., & O'Malley, C. (1996). The Evolution of Research on Collaborative Learning. In P. Reiman & H. Spada (Eds.), *Learning in Humans and Machine: Towards an interdisciplinary learning science*. Elsevier.
Dietrich, A. & Kanso, R. (2010). A review of EEG, ERP and neuroimaging studies of creativity and insight. *Psychological Bulletin, 136*(5), 822–848.
Kounios, J. & Beeman, M. (2014). The Cognitive Neuroscience of Insight. *Annual Review, 65*, 71–93. https://doi.org/10.1146/annurev-psych-010213-115154
Lambropoulos, N., P. Kampylis, S. Papadimitriou, M. Vivitsou, A. Gkikas, N. Minaoglou, N., & Konetas D. (2008). Collaborative Creativity and Hybrid Synergy for Virtual Knowledge Working: the joy of cooperation, collaborative creativity, and imagination. In J. Salmons & L. Wilson (Eds.), *Handbook of Research on Electronic Collaboration and Organizational Synergy* (pp.83–102). IGI Global Publications.
Lambropoulos, N. & Romero, M. (2015). *21st Century Lifelong Creative Learning: A Matrix of Innovative Methods and New Technologies for Individual, Team and Community Skills and Competencies*. Nova Publishers.
Martindale, C. (1990). *The Clockwork Muse: The Predictability of Artistic Change*. Basic Books.
Piaget, J. (1988). Education for Democracy. In K. Jervis & A. Tobier (Eds.), *Education for democracy: Proceedings from the Cambridge School conference on progressive education* (pp. 1–9). The Cambridge School.
Repenning, A. (2012). Programming Goes to School. *CACM, 55*(5), 38–40. https://doi.org/10.1145/2160718.2160729
Robinson, K. (2009). *Finding Your Element. How Finding Your Passion Changes Everything*. Penguin Group USA.

UNESCO (n.d.). *Technology & Learning definitions: Collaborative Learning.* http://www.unesco.org/education/educprog/lwf/doc/portfolio/definitions.htm
Vygotsky, L. S. (1978). *Mind in society: The development of higher psychological processes.* Harvard University Press.
Wallas, G. (1926). *The Art of Thought.* Franklin Watts.
Students' Innovation on Scratch and YouTube:
- 6dimpat. (2015, March 27). *Το Σπίτι του Μυστηρίου* [The Mystery House] [Video]. Scratch. https://scratch.mit.edu/projects/54660312/
- Niki Lampropoulou. (2015a, March 6). *Lancer Evolution Robot, Δ Τάξη στο 6ο Δημοτικό Σχολείο Πάτρας* [Lancer Evolution Robot, Year 4 students at the 6th State Primary School of Patras] [Video]. Youtube. http://youtu.be/so0AgEHwVxQ
- Niki Lampropoulou. (2015b, March 6). *Robotokton, Δ Τάξη στο 6ο Δημοτικό Σχολείο Πάτρας* [Robotokton, Year 4 students at the 6th State Primary School of Patras] [Video]. Youtube. http://youtu.be/sgyvJ42O9bc
- Niki Lampropoulou. (2015c, March 6). *Ο Δούρειος Ίππος, Δ Τάξη στο 6ο Δημοτικό Σχολείο Πάτρας* [Tojan Hource, Year 4 students at the 6th State Primary School of Patras] [Video]. Youtube. http://youtu.be/nzsrRpJAbsA
- Niki Lampropoulou. (2015d, March 6). *Ιπτάμενη Πόλη, Δ Τάξη στο 6ο Δημοτικό Σχολείο Πάτρας* [Flying City, Year 4 students at the 6th State Primary School of Patras]. [Video] Youtube. http://youtu.be/WZ8wPE3ePMc
- Niki Lampropoulou. (2015e, March 6). *Μηχανόζωο, Δ Τάξη στο 6ο Δημοτικό Σχολείο Πάτρας* [Mechanic Animal, Year 4 students at the 6th State Primary School of Patras] [Video]. Youtube. http://youtu.be/KUkh4hmboAM

Chapter 12

Students' Reasoning with the Pantograph

Anna Athanasopoulou[1], Michelle Stephan[1], & David Pugalee[1]

[1] *UNC Charlotte Center for STEM Education*

Abstract: *In this article, we present the results of a pilot research study we conducted on how the use of a pantograph promotes seventh grade students' mathematical reasoning and argumentation on proportional relationships. Particularly, the analysis of the data indicates students' ability to figure out how a pantograph enlarges or shrinks shapes in a certain scale factor, without having this knowledge before. Also, they found the relation between the scale factor and perimeter and area of rectangles. They combined their knowledge how to calculate perimeter and area of a rectangle and the measurements of length and width of the original rectangle and its image they produced with the pantograph using a different scale factor. In terms of the angles, they estimated that they are preserved and then they verified that they are right angles. Students achieved this goal solving appropriate activities researchers designed and used to guide students' thinking process indirectly.*

Keywords: *Geometry, reasoning, tools, argumentation*

Euclidean geometry and geometric investigations have been a central focus in mathematics from the time of classical Greek culture. According to Healy and Hoyles (2002), students have difficulty with mathematical argumentation as it relates to geometric concepts. Vincent (2002) reports that the pantograph may allow students to explore geometric ideas related to proportional relationships. Mathematical tools allow students to connect figures or drawings to visualize how the process produces related results (through scale). This shows an understanding of the characteristics of proportional reasoning according to TexTeams (Shechtman et al., 2006). The pantograph embodies mathematical properties and relationships as to allow the geometrical transformation, such as, symmetry, reflection, translation and homothety. Students using such tools develop technical skills that parallel the kinds of reasoning used in the workplace. Leak et al. (2017) found that topics and tools relate to the curriculum of physicists, engineers, and technicians and can be modified to emphasize mathematical topics and tools needed for the 21st century workplace. This study will focus on how the use of such tools promotes mathematical reasoning.

1. Study Design

1.1. Purpose of the Study

This pilot study was conducted in order to determine the viability of using the pantograph as a tool to support 12–15-year-old students' geometric reasoning. Figure 1 shows a picture of a pantograph, a wooden or plastic device that students can manipulate to draw an enlarged (or smaller) image given an original shape. The three co-authors designed activities for four students that attended an after school program and agreed to work with us for six, 45 minute sessions. The research questions that guided our design, implementation and analysis of the data were:

> 1. How do students use argumentation as they engage in work with the pantograph?
>
> 2. How does the use of the pantograph support students' ideas about proportional relationships?

The four students worked in two teams, paired according to similar abilities. The sessions fit the characteristics of a teaching experiment articulated by Steffe and Thompson (2000) that is used so that researchers can experience, firsthand, students' mathematical learning and reasoning. The videotapes were transcribed and pseudonyms used to protect the students' identity. Student artifacts were copied to add to the transcription analysis. The data were analyzed using the constant comparison method of Strauss and Corbin (1990). Themes were identified and narrative and exemplars used to develop characterizations of reasoning as students used the pantograph.

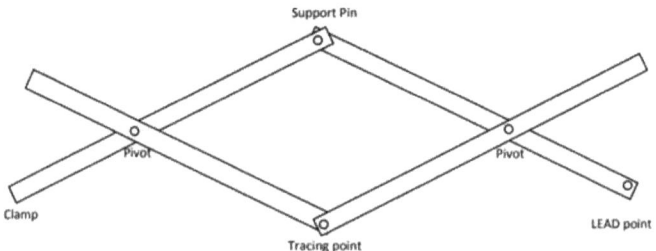

Figure 1: Pantograph

1.2. Participants

Four 7th grade students participated in this research, Anders, Natania, Navid, and Eamon. Anders was currently enrolled in a mathematics course two levels above grade while the other three students were enrolled at grade level. Anders and Natania worked as team for a total of four sessions. Anders completed two extra sessions, one by himself due to Natania's absence and one in which he joined Navid and Eamon. Navid and Eamon worked together over seven sessions. Due to space constraints, we were only able to describe the reasoning of two students. We chose Natania and Navid because their reasoning was different than their peers and provides interesting findings. This paper provides narratives related to Notions of Scale Factor and Perimeter Change. Ideas of Changes in Angles and Changes in Area were also investigated in the study.

2. Findings

2.1. Natania

Natania decided to discontinue sessions after the fourth one. However, her contributions during those sessions indicate that she knew very little about scale factor and dilations prior to the project but developed a multiplicative interpretation at the end, although unstable. In the first and second sessions, notably Natania brought the surgical video context into the discussion as a realistic context for magnification (e.g., "just like the picture [in the video], it got magnified x 2").

2.1.1. Notions of Scale Factor

Natania was unaware of the term scale factor until introduced by Anders. This conversation occurred when they allowed to explore the pantograph by tracing along a straight line on the paper:

>Natania: It's just like a heart tracker (EKG).

>Anders: It translates it! It duplicates it! It amplifies! Looks like it's making it larger! And that's the scale factor (points to the number on the pantograph).

>Natania: That makes sense!

>Researcher: Are you sure it's 2?

>Natania: We can measure it (gets a ruler). Should we use inches or cm?

Researcher: Are you sure it's 2?

Natania: We can measure it (grabs a ruler). Two inches. Here it is 4. I was right.

This dialogue indicates that although Anders had never used the pantograph, he was excited to discover that it dilates a given figure and even uses the term scale factor to describe the growth. Natania, for her part, is unfamiliar with dilations and scale factors, and, when challenged, decides to measure to determine whether the line indeed doubled its length.

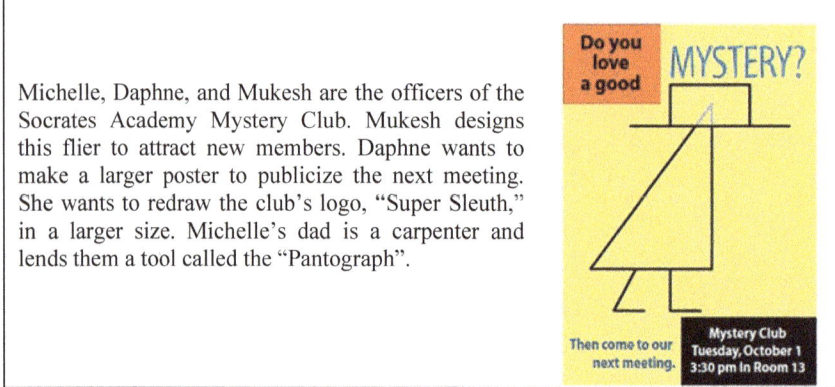

Figure 2: Mystery Figure Activity

Natania's understanding of scale factor is unstable as they continue to solve problems. When introduced to the Mystery Club problem (Figure 1), they create a poster 10 times larger; however, with a large-scale factor, the pantograph is limited in movement. Natania claims that the new poster would look like the long, thin rectangle in the middle of the pantograph seen in Figure 2, indicating she does not yet see the scale factors as stretching all sides by the same amount.

Figure 3: Natania's "poster" in the middle of the pantograph

As the sessions continued, Natania's understanding of the impact of the scale factor on the original figure was that it doubled, tripled (multiplied by) the lengths of the sides to create the image. However, when a decimal scale factor was used, Natania's understanding was shaken. When presented with an original Super Sleuth (the shape in Figure 1 above) and an image created by a 150% enlargement, Natania immediately used a ruler. When she measured the original hat length, she determined it was 6cm and the image was 9cm for a scale factor of "3 times." Natania reverted back to additive reasoning when confronted with a decimal scale factor.

2.1.2. Perimeter Change

During the second session, when presented with the Super Sleuth problem, Anders and Natania created an image using a scale factor of 6 on the pantograph. The researcher asks the students to explore how the perimeter change, if any at all.

> Researcher: What's your first intuition?
>
> Natania: My first to get a ruler to see if...
>
> Researcher: What's your first guess? How her perimeter compares to super sleuth's.
>
> Natania: It's gonna be, well it seems bigger.
>
> Researcher: It's definitely bigger. OK. It's gonna take more lead over there.

> Natania: Well, since we said it was going to be about 6 times as bigger, it might take 6 times more?

Natania, again grounds her intuition first in the physical act of measuring (*get a ruler to see...*), but, when pushed, conjectures that the perimeter may be 6 times larger since the pantograph is set on 6. Anders, for his part, argues that the perimeter will be 12 times larger (double the scale factor). Upon measuring, they discover that Natania's conjecture was correct.

2.1.3. Summary

In summary, for Natania, the pantograph was used as a tool for creating an image rather than a reasoning device. The ruler was a more dependable tool for exploring the relationships between the original and image figure. In fact, Natania repeatedly used the ruler to verify intuitions or uncertainties and did not provide mathematical reasons for her findings. For example, to justify why the perimeter should increase by the scale factor, she simply said that the shape got larger by the scale factor rather than coordinating the side length growth to the perimeter growth. The same can be said for angle measurement and area; Natania did not come to a stable understanding of those changes; relied on measuring as a way to verify these relationships and attempted to remember Anders' claims rather than understanding why they are true mathematically.

2.2. Navid

Navid attended all seven After School sessions. His contributions during those sessions indicate that he did not know anything about scale factor and dilations prior to the project but developed a strong understanding and interpretation of these concepts at the end without naming them. In the first session, Navid commented that the robot in the surgical video must be precise.

2.2.1. Notions of Scale Factor

During the first session, Navid tried to draw a square and together with Eamon tried to trace it and produce a new one using the pantograph, for first time, without success. Then, Navid drew a line segment equal to 10cm, traced it with Eamon using the pantograph, and produced a new line segment equal to 20cm. He observed that the line segment doubled, see the dialogue below:

> Researcher: Do you think it will be easier if we start with something simpler?
>
> Navid: like a line

> Researcher: How long is the line? Could you measure it?
>
> Navid: This here is 10cm (after measuring it using the ruler)
>
> Researcher: And this here?
>
> Navid: It is 20cm.
>
> Researcher: So, what happened?
>
> Navid: It doubled.

Navid's understanding of scale factor, and how the pantograph creates it, becomes clear as they continue to solve problems. When introduced to the Mystery club problem (see Figure 1), they first make the poster two times bigger and then three times bigger. Navid measures the original sides of the rectangle and of the right triangle, conjecturing that corresponding lengths in the produced picture are to scale, and measures to verify his conjecture. He was able to observe the points that move and the stable point in the pantograph, during the tracing-drawing process.

When presented with an original Super Sleuth (the shape in Figure 1 above) and an image created by a 150% enlargement, Navid measures the original hat length (4 cm) and the image and determines that the image (about 6 cm) is one and half times bigger.

> Navid: So, this (he points at the image) is almost one and a half bigger than that one (he points at the original). That's why we say 100 and 50.

After these calculations, they were able to set up the pantograph on 1½, to trace the original, to produce the new image, and by measuring Navid verified that the new image is 1½ times bigger.

> Researcher: Navid explain to Eamon what you measured (during the measurements Eamon was out of the room to drink water)
>
> Navid: This one I took the ruler and I measured it and I got 3cm (the shorter side of the right triangle) so the same thing with this one and then I got 4.4cm so, it is about 1½ times more.

Navid's determination of the enlargement by 1½ times indicates the development of his understanding of the impact of the scale factor on the original, using the pantograph. The next feature Navid discovers is how pantographs produce images in scale. Navid and Eamon set up three pantographs one on 1½, the second on 2, and the third on 3, where in the middle a rectangle is created. By measuring the sides of these rectangles, Navid observed that the lengths of sides of the rectangles on 1½ and on 3 are equal the other way around. Focusing on the stable point, the tracing points, and the drawing point, Navid decides to measure these distances. He found that the distance from the stable point to the tracing is 14cm and from the tracing point to the drawing the distance is 7cm. Then, he observed that 14 plus 7 is 21 and 21 to 14 is 3 to 2 equal to 1½. Therefore, he conceptualized that the ratio of the distance of the stable point from the drawing to the distance of the stable point from the tracing point is 1½, creating the enlargement of the image by 1½ times.

After the conceptualization of enlargement of shapes using a pantograph, the next question was how we could shrink the original shape into ½. Anders also joined this session and he set up his pantograph on 1½, Navid set up his pantograph on 2, and Eamon set up his pantograph on 3 and they traced the original picture. Observing all three images, Navid observed that the 1½ times bigger image is half the three times bigger, but the question was how we could draw an image ½ the original with one pantograph. Anders observed that by switching the lead with the tracking point, the image shrinks and if the pantograph is on 2 we draw an image half of the original.

>Anders: You switch the tracking point in the lead tip and to then when you do that but after you do that you will be able to trace this and get into that.

>Researcher: And then how can we make it half?

>Anders: You set it to 2

Understanding what Anders explained, all of them switched the lead with the tracking point and traced another copy of the original drawing an image half the original. Then, Navid conceptualized the process to make half of the original.

>Researcher: What distance makes the half?

>Navid: Put all of them on 2 and then switch these two.

The last shape Navid and Eamon worked with was a circle. Navid chose to use a scale factor of 6. When he traced the circle, the hard part was what to measure to verify that the new circle is 6 times bigger the original. Navid did not know any of the characteristics of a circle.

>Researcher: Do you know any piece on the circle with a name?

>Navid: Doesn't it have like something that is called circumference or something?

> Researcher: What is circumference?
>
> Nick: I had never it is, but I just have heart of things like circumference radius
>
> Researcher: So, you know radius, you know circumference.
>
> Nick: Well I have heart of them, but I have no clue what they are.

Then, the researcher helped Navid and Eamon to understand the meaning of circumference of a circle decomposing the word. To conceptualize the diameter and the radius of a circle, they folded the provided paper circle under the guidance of the researcher. Finally, Navid measured the diameter of the original circle and of its image and calculated the length of each radius by dividing by two. So, he verified that the new circle is 6 time bigger the original.

2.2.2. Perimeter Change

During the second session (Super Sleuth), Navid and Eamon created an image using the default scale factor of 2 on the pantograph. The researcher asks Navid to calculate the perimeter of the original rectangle and Eamon to calculate the perimeter of the image. Navid knew the concept of perimeter and figured out that it is 6 cm although Eamon did not have a clear understanding of the concept of perimeter. Navid helped Eamon to calculate the perimeter of the image of the rectangle. Navid using the calculations figured out that the perimeter becomes double in this case.

> Researcher: Do you know how much is the perimeter of this rectangle (the original one)?
>
> Nick: how much is the length? It is 2 so 2+1+2+1 so it is 6cm (he writes it down).
>
> Researcher: Eamon how much is the perimeter of this one (the image)?
>
> Eamon: The one side is 2 and the other 4
>
> Researcher: So, what is the perimeter?
>
> Eamon: we are times in?
>
> Navid: no the perimeter is when you add all sides, 4 and

Eamon: 10 Students write their statements at the prompting of the researcher.

Researcher: So, what do you think is the relationship of these perimeters?

Navid: Since this (points at the original picture) no sorry this was two times bigger (points at the image picture), so these forms show that the numbers are two times bigger of each other.

In the third session, Navid generalized the rule about perimeters saying that when the pantograph is on 2 it doubles, of 3 it triples, and on 4 it is 4 times bigger than the perimeter of the original.

2.2.3. Summary

In summary, for Navid, the pantograph was not only as a tool for creating an image in scale but also a reasoning device since he figured out how the pantograph works. Navid used the ruler as a tool to measure line segments and expressed mathematical reasoning applying these measurements in math relations. For example, to justify why the perimeter should increase by the scale factor he applied the formulas from his prior knowledge. Similarly, he figured out the relations of areas and the scale factor squared. At the end, Navid also conceptualized the mechanism of pantographs. He described clearly where we need to put the original shape to enlarge an image twice, three times, four times, and 1½ as well to shrink an image by ½. He developed a confidence about the use of a pantograph to create similar shapes without even knowing or to have ever heart the work "similar". In reality, Navid developed the concept of similarity through his experience with pantographs.

3. Conclusion

This teaching experiment explored how four 12–15-year-old students thought about proportional relationships as they used a pantograph. Students' use of argumentation was a key focus. Data was analyzed using a constant comparative method. Themes provided characterizations of students' thinking. Findings indicate that the pantograph can be used effectively as a reasoning device though students may only view the tool as a means for creating an image. Data also indicate that students can use the pantograph to explore measurements in mathematical relationships. Student explorations also demonstrate that the pantograph may facilitate the development of ideas around similarity without formal development of this concept. The teaching experiment also underscored how additional scaffolding is necessary with some students to bridge their procedural views of relationships with conceptual potentially developed through the use of mathematical tools.

REFERENCES

Healy, L., & Hoyles, C. (2002). Software tools for geometrical problem solving: Potentials and pitfalls. *International Journal of Computers for Mathematical Learning*, 6(3), 235–256. https://doi.org/10.1023/A:1013305627916

Leak, A. E., Rothwell, S. L., Olivera, J., Zwickl, B., Vosburg, J., & Martin, K. N. (2017). Examining problem solving in physics-intensive Ph. D. research. *Physical Review Physics Education Research*, 13(2), 020101. https://doi.org/10.1103/PhysRevPhysEducRes.13.020101

Shechtman, N., Knudsen, J., Roschelle, J., Haertel, G., Gallagher, L., Rafanan, K., & Vahey, P. (2006, April 7–11). *Measuring Middle-School Teachers' Mathematical Knowledge for Teaching Rate and Proportionality* [Paper presentation]. Annual meeting of the American Educational Research Association (2006 AERA). San Francisco, CA, USA.

Steffe, L. P., & Thompson, P. W. (2000). Teaching experiment methodology: Underlying principles and essential elements. In R. Lesh & A. E. Kelly (Eds.), *Research design in mathematics and science education* (pp. 267– 307). Erlbaum.

Strauss, A., & Corbin, J. M. (1990). *Basics of qualitative research: Grounded theory procedures and techniques*. Sage Publications, Inc.

Vincent, J. (2002). Dynamic Geometry Software and Mechanical Linkages. In D. Watson, & J. Andersen (Eds.), *Networking the Learner: Computers in Education. 7th IFIP World Conference on Computers in Education-WCCE 2001* (pp. 423–432). Springer.

Chapter 13

Introduction to Physical Computing with Raspberry Pi in a STEM Education Framework

Nikolaos Balaouras[1], Anthi Karatrantou[2], Georgios Panetas[2], & Christos Panagiotakopoulos[2]

[1]*School of Pedagogical and Technological Education (ASPAITE) Branch of Patras*

[2]*Department of Primary Education, University of Patras*

Abstract: *STEM education refers to the scientific approach of solving a problem using tools from Science, Technology, Engineering and Mathematics. This multidisciplinary method allows students to develop a variety of skills needed for their transformation into integrated members of a society where everything is interconnected and continuously is developing technologically. In this paper an attempt was made to introduce students of 9th-grade to the world of STEM education through physical computing using the Raspberry Pi platform and the Python programming language. Students were asked to work in groups to design, construct and program a circuit (system) that simulates the function of traffic lights for visually impaired people. One group of schoolteachers also, attending a STEM postgraduate course, worked with the same tools and the same educational activities in order to discuss their usefulness and applicability in the classroom. Analyzing students' work as well as schoolteachers' thoughts, useful information can be derived on how physical computing can support the aims of STEM education, as well as on the benefit of such educational activities.*

Keywords: *Physical Computing, Raspberry Pi, Python, STEM Education, Automation Systems*

1. Introduction

In a rapidly developing technologically world which tends to be interconnected, a strong need is arising to educate the new members of society with new, innovative tools and methods that will use in practice concepts, tools, procedures and phenomena from more than one science in order to understand in depth the real things around them. STEM education can cover this need, offering students skills of Science, Technology, Engineering and Mathematics. In this context, educational robotics, automated control systems and physical computing provide students with the opportunity to use STEM methods, since they will need to combine and apply knowledge from the four pillars of the method (Physics, Technology, Engineering Science, Mathematics), to solve authentic problems of the modern world (He et al., 2016). Physical computing can effectively serve this framework as it is related to the design and construction of interactive physical objects and can support students to bridge the gap between virtual and real world developing in this way a basis for them to redefine their knowledge (Psycharis, 2015).

In this paper an attempt was made to introduce physical computing and STEM methodology to students of Secondary Education. Students worked in groups in order to create a practical and social application combining knowledge from Computer Science, Technology and Science courses attended in Junior High School. On the other hand, schoolteachers attending a postgraduate course in a STEM Education program, worked using the same tools and completed educational activities in order to discuss their usefulness and applicability in the classroom.

2. STEM Education and Physical Computing

Nowadays the rapid growth of technology, led many countries to shape their education system to prepare the new citizens for new methods of exploration and problem solving. The European School Network (*www.eun.org*) since 2009 is moving to this direction by piloting in schools new technologies and learning activities exploring the use of new pedagogical tools for STEM education. STEM education can transform the traditional teacher-centered approach to constructive teaching and learning by equipping students with the critical 21st century skills such as teamwork, problem solving, project management, analytical and synthetic thinking, creativity (He et al., 2016).

STEM education is based on the idea of educating students in Physics, Technology, Engineering and Mathematics in a combined way drawing knowledge and tools from each science in order to solve an authentic problem. Educational activities serving STEM methodology are continuously designed and developed, mainly by creating and programming robotic applications and simple automated control systems. Today, a large variety of kits is available with the Lego Mindstorms (RCX, NXT, EV3), the Arduino platform, and the Raspberry Pi platform to be more and more popular (Balaouras, 2018).

Physical computing combines digital and analogue elements and is a perfect tool for STEM applications as it "*is about creating a conversation between the physical world and the virtual world of the computer*" (Eguchi, 2014, pp. xix). It gives the opportunity for very interesting and creative projects as it is linked to that part of Computer Science that allows the creation of "art" within the classrooms. It essentially allows students to create tangible applications with usefulness in the real world and society acquiring by this way useful skills and experiences, even in their professional orientation. The term was first mentioned by O'Sullivan and Igoe (2004) who saw it as a crucial element of the systems which use transducers (sensors and actuators) to connect the virtual and physical world (Przybylla & Romeike, 2014).

According to Przybylla and Romeike (2014) there are three basic pillars of physical computing. *Processes*, *products* and *tools*. In *processes*, the developer is called upon to design processes and applications that connect the digital world with the physical one, unifying hardware and software. O'Sullivan and Igoe (2004) encourage creators to forget what they knew about computers and urge them to work on the needs of society, people and the environment by focusing on ideas rather than on technical constraints. The second pillar is the *products*. Creators create programmable tangible media (or physical objects) that communicate with the environment using transducers and can interact with each other by creating a network of applications. The third pillar is the *tools* that developers use, such as programmable games (e.g., Lego Mindstorms) or microprocessors (Raspberry Pi).

The creative dimension of physical computing fits perfectly with constructivism in education. According to the constructionist learning theory of Papert and Harel (1991), learning is most effective when learners construct knowledge and develop competencies from their own initiative and for personally relevant purpose. Physical computing can provide the necessary tools for students in STEM education, developing skills mainly in Science, Mathematics, Engineering and Computer Science, giving them motivation for learning in the school environment (Psycharis, 2015). The main research question of this study was to explore whether physical computing can support the aims of STEM education and the benefit of this kind of educational activities.

3. Raspberry Pi

In 2008, through the collaboration of Pete Lomas and David Braben, the Raspberry Pi Foundation in the United Kingdom was created to promote the teaching of Computer Science in schools and developing countries.

Figure 1: Raspberry Pi 3 model B

Rasberry Pi (*https://www.raspberrypi.org*) is a full computer with a size of a credit card. It is an open-ended platform with a free operating system that provides many features to the users, similar to these of a personal computer. Its operating system is LINUX based. It has preinstalled programming environments that make it ideal for language programming learning such as Python, C, C ++, JAVA and Scratch for youth (Balaouras, 2018). Raspberry Pi can interact with the environment and can be used in a great number of digital creation projects as it has pins for input and output data (General Purpose Input Output pins GPIO pins), that the user can control via programming. Sensors, LEDs, motors, buttons etc., can be connected to these pins (Nuttall, 2017). The ability for connection to peripheral devices makes Raspberry Pi a fully operating small size computer and the ability for internet connectivity makes it available for any kind of upgrade (*https://projects.raspberrypi.org/en/projects/physical-computing*).

4. Python for Raspberry Pi

Raspberry Pi provides a host of programming environments whereby the user can create all kinds of applications according to his level. Pre-installed with the Raspbian operating system, the user can find programming environments for Scratch, Python, C, C ++ and Java, but there is also the ability to install any programming language environment.

The programming language used for the activity is Python (*https://www.python.org/*). It is a high-level programming language and was used because of its ease for beginners to program. Its syntax is easy and allows developers to use fewer lines of code than C or Java. It uses ready-made libraries that the developer can enter into his/her program and greatly facilitate that the code does not have many lines, to be easy to read and understand, and perform operations quickly and easily (Balaouras, 2018). The library used with the activity was the "*gpiozero*" that provides component interfaces to allow a frictionless way to get started with physical computing (Nuttall, 2017).

5. Methodology

The aim of this study was to introduce physical computing and STEM methodology to students of Secondary Education. Students called to work in groups in order to create a practical application using Raspberry Pi and Python combining knowledge from Computer Science, Technology and Science. Schoolteachers attending a postgraduate course in a STEM Education program worked also, using the same tools and completed the same educational activities, in order to discuss their usefulness and applicability in the classroom.

5.1. Students

The research took place in a public high school in Patras. A total of twenty (20) 9th grade students (eight (8) girls and twelve (12) boys) participated voluntarily in the study. The students were divided randomly into five (5) groups of four (4) members and the educational activity lasted two 3-hour sessions in the Computers Laboratory of the school. Students had coped with the widest range of subjects and have been taught about electricity, circuits and their elements (sources, lamps, switches and resistors), and their basic properties. They were familiar with the Logo programming environment and have used basic programming structures in simple programs. The whole educational activity was based on four (4) Worksheets of increasing difficulty that supported students to gradually construct and program more complex circuits aiming to crate finally a system of traffic lights for visually impaired people.

At the end of the procedure students were expected to *recall and make use of knowledge from their science subjects, recognize basic electronic circuit elements, be able to explain how a simple control system works, use basic programming structures and choose the appropriate programming structures to control their construction, to be familiar with new tools as Raspberry Pi and Python, realize how Engineering, Science, Mathematics and Technology are connected, recognize applications of physical computing in everyday life, adopt positive attitude for innovation in everyday life.*

Four data collection methods were used: monitoring (groups and members) and personal notes of the researchers, analyzing answered worksheets by the students while working, answering of short questionnaires before (Diagnostic Questionnaire) and after the activity (Evaluation Questionnaire) and a short semi-conducted group interview. Two teachers-researchers were observing the discussions, activities and reactions among the students. They kept notes and made interventions when students needed help, adopting a supportive and facilitative role of students' work and learning. The Diagnostic Questionnaire consisted of ten (10) questions (closed-type and opened-type with a short answer) and students had to give anonymous answers about their gender, their ability to use computers and internet, their previous programming knowledge and experience, their knowledge about automatic control systems and physical computing. The Evaluation Questionnaire was used to evaluate the overall procedure (tools, activities, learning outcomes) and consisted of nine (9) questions (closed-type and opened-type with a short answer).

During the 1st session the Diagnostic Questionnaire was answered and a familiarization phase with Raspberry Pi, the basic elements of a circuit (breadboard, resistors, LED, switches, etc.) and Python took place. Students worked based on the 1st and 2nd worksheet according to which they had to construct simple circuits with a LED, resistors, a switch and a buzzer. They used simple programming structures (*sequence*, control structure *While loop*) to program and control the function of the circuits. During the 2nd session students worked on the basic activity based on the 3rd and 4th worksheet. They called to design, construct and program a circuit (system) that simulates the function of traffic lights for visually impaired people. They used programming structures such as *sequence*, control structures *While loop* and *If...Then...Else*, the command *Random* and simple subprograms to program and control the function of the final circuit. At the end, a discussion about the activity, a short discussion about applications of simple automatic control systems and physical computing in everyday life as well as the evaluation of the whole procedure (Evaluation questionnaire) also took place.

5.2. Schoolteachers

Five (5) schoolteachers (four (4) women and one (1) man) also, attending postgraduate courses in a STEM Education program worked with the same tools and educational activities in a five (5) hours workshop. Schoolteachers were supported by the researchers to fulfill the activities especially on programming the circuits due to their lack of previous experience in programming. At the end of the procedure they answered a questionnaire consisted of eight (8) opened-type questions in order to write down their opinions about the usefulness and applicability in the classroom as well as the added educational value of such educational activities.

6. Findings and Discussion

6.1. Students

According to the students' answers to the diagnostic questionnaire all students had partial knowledge on the use of computers and internet, had attended lessons on

LOGO and Scratch programming at school and could incompletely describe the control structures but none of them knew about electronic circuits, automated systems and physical computing.

According the completed worksheets and the researchers' notes it seemed that all students worked in groups discussing and interacting each other. Each group followed step by step the guidelines of the worksheets to create and program the circuits. They could easily recognize the electronic elements to use and how they work but it was difficult for them to put the elements on the breadboard and connect them correctly. It was difficult for them to recognize the role of the *Ground* and *Voltage* in a circuit, possibly due to the lack of the corresponding practical knowledge coming from Science.

 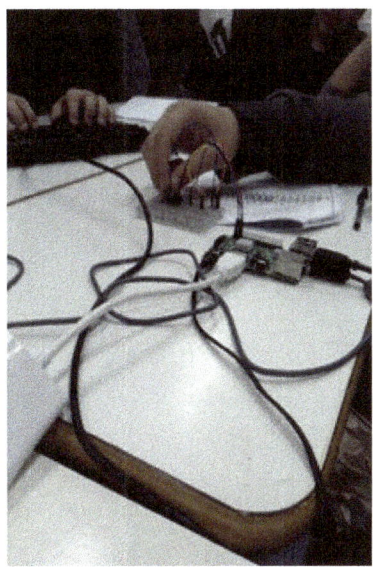

Figure 2. (a) Preparing the material   (b) Constructing the circuit

They needed support and help to program their circuits because of their limited knowledge and experience in programming. They used Python for first time so they faced difficulties to use libraries and write the commands but they could easily designthe algorithm and choose the appropriate programming structures each time. They were excited each time their circuit was functivoning as they had programed it and they liked to make experiments on it. It was the first time for them to use subprograms so the researchers had to explain the role of a subprogram and the benefits of using it in a program. All groups worked calmly with discussions, arguments, agreements and disagreements and their interest was important during the whole procedure. Each member tried to contribute to the activity, there was a work and responsibilities sharing among members and leaders were appeared in most groups. In all groups, the discussion among members took place to find solutions and complete their task each time although the limited time they had to work caused difficulties so as to think deeply on new things they met.

Figure 3: (a) Students in action   (b) Final circuit

According to the students' answers at the evaluation questionnaire, after the end of the whole procedure all students were able to briefly describe the electronic elements they used (LED, resistors, switch, buzzer) and their role in the circuit. Eight (8) students faced difficulties to describe the role of the While loop in their programs although they used it properly during working. All of them expressed their pleasure because for first time they created something from scratch and watched it working. A typical answer was: «*It is the first time I worked with my fellow students and created many things that at the beginning I thought that I could not create them*». Sixteen (16) students expressed their particular expression about the Raspberry Pi platform and the way it can be connected with circuits, systems and real things. Eighteen (18) students would like to work on such activities again and learn more things about automated systems and physical computing. Eighteen (18) students seem to realize the practical value of such applications in everyday real life and could recognize such applications in real physical world. Students rated their liking for the activities with a mean rate of 8 and the usefulness of the activities with a mean rate of 7 (rating scale 1-10).

6.2. Schoolteachers

According to the schoolteachers' answers and opinions the educational added value of such tools and activities seems to be important as they introduce students to new concepts and applications with a multidisciplinary way keeping them continuously active. A characteristic answer was «*these tools and activities support students to be activated, to try, to correct, to take initiatives and to suggest solutions...*». As basic advantages of Raspberry Pi they reported the openness of the hardware and the software as well as its function as a cheap full computer of extra small size. Schoolteachers supported that the connection between students' knowledge and modern technology requirements and needs using interaction, critical thinking, exploiting mistakes and creating a team spirit can be achieved via Raspberry Pi applications and physical computing. Schoolteachers expressed their willing to use these tools in their classroom as they believe that their students can gain knowledge concerning Programming, Science, Mathematics, Technology, Engineering moving from theory to practice within a playful way. At the same time, all of them noticed that schools and teachers need an appropriate budget to buy, maintain and use this tools in their classroom. They also noticed that curricula should be redesigned to include such activities in an integrated framework. All of them would like to be trained in these field, although they faced difficulties. A typical answer express their thoughts: «*During the activities we escaped from the classical lesson, experimented and learnt in a creative way and a fun and pleasant climate developed among us as well...*».

7. Conclusions

In this study, an attempt was made to introduce physical computing and Raspberry Pi to students of secondary education in a STEM education framework. Students worked in groups to design, construct and program system that simulates the function of traffic lights for visually impaired people. Schoolteachers attending a postgraduate course in a STEM Education program worked also, using the same tools and completed the same educational activities, in order to discuss their usefulness and applicability in the classroom.

Students during their work faced difficulties but they managed to solve problems arose with the help of their schoolmates and the researchers. They faced difficulties with the creation of the circuits and they needed to recall knowledge from Science. They also faced difficulties using Python and they recalled knowledge from algorithms, LOGO and Scratch programming. There was an increasing feeling of success as they were fulfilling the activities of each worksheet and could enjoy the results of their work. Almost all students liked the educational activities and recognized their usefulness in everyday life.

Schoolteachers involved in the activities recognized an educational benefit of this kind of tools and applications but they argued for support from the educational authorities at the technical support level and the training level as well. All of them could use such activities in their classroom after appropriate training.

Concluding, physical computing and STEM methodology can offer important learning opportunities to students of any level. However, the prerequisites for that include teachers widely training in these fields with the appropriate support and budget. Many projects at school should be developed, implemented and evaluated, to derive safe results and conclusions. This paper aims to contribute to the wide discussion among educators and authorities for a better, authentic education and training of young people according to the 21st century needed skills.

REFERENCES

Balaouras, N. (2018). *Utilization of Raspberry Pi Platform for introducing basic principles of automatic systems to students in Secondary Education* [Unpublished doctoral dissertation]. School of Pedagogical and Technological Education, Patras.

Bayer Corporation (2016). *Planting the Seeds for a Diverse U.S. STEM Pipeline: A Compendium of Best Practice K-12 STEM Education Programs*. MSMS-Resources. https://www.makingsciencemakesense.com/static/documents/Resources/K-12-STEM-edu-programs.pdf

Eguchi, A. (2014). Educational Robotics for Promoting 21st Century Skills. *Journal of Automation, Mobile Robotics & Intelligent Systems*, *8*(1), 5–11. https://doi.org/10.14313/jamris_1-2014/1

He, J., Lo, D. C.-T., Xie, Y., and Lartigue, J. (2016, October 12–15). Integrating Internet of Things (IoT) into STEM undergraduate education: Case study of a modern technology infused courseware for embedded system course [Paper presentation]. In IEEE (Ed.), *Proceedings 2016 IEEE Frontiers in Education Conference* (pp. 1–9). https://doi.org/10.1109/FIE.2016.7757458

Karatrantou, A., Panagiotakopoulos, C. (2012). Educational Robotics and Teaching Introductory Programming Within an Interdisciplinary Framework. In A. Jimoyiannis (Eds.), *Research on e-Learning and ICT in Education* (pp. 195–208). Springer Science and Business.

National Research Council (2011). *Successful K-12 STEM Education: Identifying Effective Approaches in Science, Technology, Engineering and Mathematics*. The National Academies Press.

Nuttall, B. (2015, November 24). Piozero Documentation. *Raspberry Pi Blog*. https://www.raspberrypi.org/blog/gpio-zero-a-friendly-python-api-for-physical-computing/#

O'Sullivan, D., & Igoe, T. (2004). *Physical computing: sensing and controlling the physical world with computers*. Course Technology Press

Papert, S., & Harel, I. (1991). Situating constructionism. In S. Papert & I. Harel (Eds.), *Constructionism* (pp. 1–11). Ablex.

Psycharis, S. (2015). The impact of computational experiment and formative assessment in inquiry-based teaching and learning approach in STEM education. *Journal of Science Education and Technology*, *25*(2), 316–326. https://doi.org/10.1007/s10956-015-9595-z

Przybylla, M., & Romeike, R. (2014). Physical Computing and its Scope Towards a Constructionist Computer Science Curriculum with Physical Computing *Informatics in Education*, *13*(2), 241–254. http://dx.doi.org/10.15388/infedu.2014.05

Schulz, S., & Pinkwart, N. (2015). Physical Computing in STEM Education. In J. Gal-Ezer, S. Sentance, J. Vahrenhold (Eds.), *Proceedings of the WiPSCE Workshop in Primary and Secondary Computing Education* (pp. 134–135). ACM.

Chapter 14

Educational Robotics & Children's Attitudes Towards STEM

Marina Karemfyllaki[1], Anthi Karatrantou[2], & Christos Panagiotakopoulos[2]

[1]*MSc STEM in Education (ASPAITE) Branch of Athens*

[2]*Department of Primary Education, University of Patras*

Abstract: *The STEM methodology in the field of education is commonly understood as an educational approach that includes Science, Technology, Engineering and Mathematics. Educational robotics could be considered as a vehicle for new ways of constructive learning and as a vehicle that leads to new learning paths that are an integral part of STEM education and culture. This research was carried out with the participation of 25 children of the fifth grade of a primary school in a suburb of Athens. The sample divided into groups of 5 members. The children in every group were asked to solve an "authentic" problem with the use of Lego EV3 educational packages. Data were collected with appropriate questionnaires. The results of the data analysis showed that children's attitudes towards STEM-related sciences were more positive and their desire to pursue a career in more than one of the future scientific areas concerning STEM was increased.*

Keywords: *STEM education, Educational robotics, Lego Mindstorms EV3*

1. Introduction

Today it is well known that the use of the STEM methodology in education can benefit from the use of educational robotic technologies. Moreover, educational robotics-based learning activities can enhance not only the acquisition of concepts in several fields (science, programming, mathematics, engineering etc.), but can also improve the emotional and social development of children. The discoveries in the fields of Engineering, Science and Technology have led society to tremendous developments in the 20th century, and scientists expect a similar trend in the upcoming decades (National Academy of Engineering, 2008). On the other hand, researchers and economists predict accelerated growth of jobs in these scientific areas. For example, the Department of Commerce of USA has already found that in the first decade of the 21st century the number of job opportunities in these sectors has increased three times faster than in other sectors (United States Department of Commerce, 2012). These new jobs will continue to require knowledge and skills in Science, Technology, Engineering and Mathematics (STEM), and an advanced set of general skills in critical thinking, communication and collaboration, often referred to as "Skills of the 21st Century". From the research have be proven that if we would like to help children to find an identity in the STEM community we must give them opportunities to conduct real science research and engineering design. Exploring different professional identities (e.g., what kind of work they would like for themselves) is an important aspect for identity formation and maturity (Malanchuk et al., 2010).

The purpose of this study is to examine whether children's attitudes towards the STEM-related sciences and their interest in pursuing a career in related fields are influenced by their interaction with educational robotics.

2. STEM in Education and Educational Robotics

Educational robotics is defined as the use of robotic constructions as a pedagogical tool (Riberio et al., 2008). Over the last two decades, the interest in the use of educational robotics in the school has increased. Educational robotics is an effective learning tool based on project-based learning in which STEM, coding, computational thinking and engineering skills are incorporated into a project (Eguchi, 2014).

The use of educational robotics provides children with opportunities to query and think deeply about technology. During designing, constructing, planning and evaluating autonomous robotic constructions, children not only learn how technology works but also apply the knowledge and the skills they learn in the school in a meaningful and exciting way (Stergiopoulou et al., 2016). Educational robotics is a tool which provides the opportunity to integrate not only STEM but also many other disciplines, including linguistics, social studies, dance, music and art while help children to find new ways to develop collaboration skills, with critical and innovative thinking. Finally, educational robotics is a learning tool that improves children experience with learning by doing (Eguchi, 2014).

The framework for integrating STEM into the classroom consists of some key elements (Johnson et al., 2016), such as: environments with personal meaning, engineering challenges, redesign based on the children's knowledge, teaching the content adapted on the way the children learn, and work in groups to increase the communication skills of the participants. The implementation of a STEM lesson is often combined with Computational Science (Psycharis, 2016), while computational thinking is used as a method to solve a problem (Wing, 2006).

3. Lego Mindstorms EV3

The Lego Mindstorms (LM) EV3 is a Lego product which belongs to the third-generation kit. It is a handheld robot that includes a wide variety of bricks, motors, sensors and other equipment that helps the construction of real models. These robots can be programmed to execute commands and to react to different stimuli received through their sensors using the appropriate programming development environments. In terms of introductory programming, the use of robots is expected to have a positive impact as it can help among other things in understanding of an accurate and logical language for the machine (Komis, 2004).

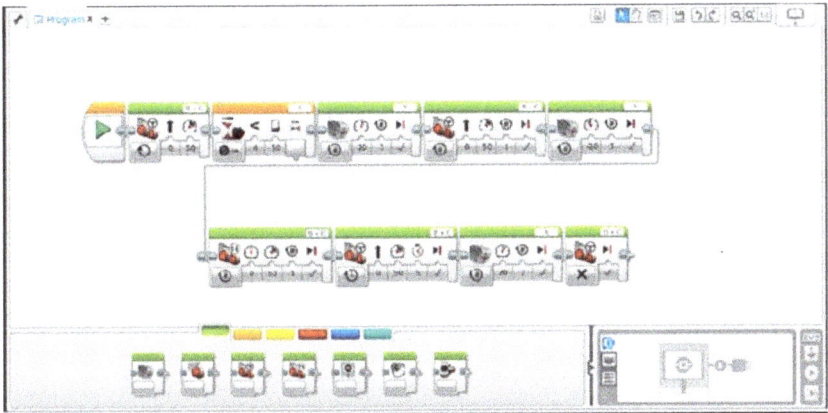

Figure 1: Programming LM EV3 (screenshot)

LMs are used as a tool to teach problem-solving methods, as it is very pleasant and interesting, while offering a simple educational environment. Children see it more as a game than as educational tool, as the majority of children have played with Lego bricks in the past. Note, that the game is a very important factor that promotes and encourages children to learn (Komis, 2004). LMs Software for programming EV3 modules is based on the use of commands in the form of icons for establishing concrete programmable "behaviors" to these mechanical constructions (Figure 1).

The child places with drag-and-drop the appropriate commands in a sequence and in the order he/she wants to be executed by the robot. When the program is completed, it is transferred to the robotic construction and the robot is ready to execute the commands acquiring the desired behavior.

4. Children's Attitudes Towards STEM

The main of the factors examined in this study is children's attitudes towards STEM methodology. In a recent survey on high school children's attitudes about Mathematics, revealed that fewer than half of the participants noted that they like Mathematics (Greenspan, 2000). The researcher connects much of this negative attitude to the children's anxiety about Mathematics, which reduces children's self-confidence in their mathematical skills. This lack of self-confidence reduces their self-esteem and their motivation. In addition, it has been documented that a major problem that repels children from the STEM-related sciences is behavioral perceptions about the relevant disciplines. For example, engineering is considered heavy for manual work, while math-based professions are only for children who are well prepared on this subject (Bowen et al., 2007). Perceptions and such attitudes cause children to choose unrelated studies and careers with these sciences.

On the other hand, several classifications have been developed to detect children's attitudes towards STEM, which can be applied in a useful way (Faber et al., 2013; Guzey et al., 2014; Tseng et al., 2013; Tyler-Wood et al., 2010). However, most of these studies were conducted in several countries (mainly in the USA), and there is no study related to Greece. The USA remains world leader in discovery and innovation today because of STEM education that is already prevalent there (Faber et al., 2013). Therefore, it is important for a country to improve children's creativity and competitiveness through STEM education (Suprapto, 2016).

5. Research Question

The research question for this study was as follows: *Do children's attitudes towards the STEM-related sciences and their interest to pursue a career in related fields are affected by their interactions with educational robots?*

Specifically, the participants should think about how to deal with an environmental problem in their area. In the process for the solution of the problem, the participants collaborated and used the Lego robotics system in a STEM education environment. LMs EV3 robotic kits were used in the study just because their use is based on the problem-centered and exploratory teaching and learning method.

6. Methodology

The sample consisted of twenty-five (25) children (12 girls, 13 boys) of the 5th grade of a public elementary school in a suburb of Athens.

The research conducted in a two-week period in the school's multi-purpose classroom using laptops and educational robotics kits, and in three (3) consecutively workshops. Every workshop lasted two (2) hours. Throughout the workshops, the researcher supported with more four (4) mentors, in order to guide and support the children's activities.

The sample was divided in five (5) groups of five (5) members, according to participants' choice in the 1st workshop, which was introductory in Lego robotics and in the following workshop they remained with the same classmates. The children did not have previous experience with the educational robots LM EV3 except two (2) of them. All children had been taught the basic Scratch programming language in their school.

6.1. Data Collection Instruments

The data were collected using the *S survey questionnaire*, developed by the Friday Institute (2012), and used to determine the effect of engagement on STEM activity on children's attitudes. In the questionnaire are included two components group of questions: (a) the measure of changes in STEM self-confidence and effectiveness in STEM-related disciplines (e.g., in Mathematics, Science, Engineering and Technology) with 26 questions, and (b) the participants' interest in pursuing careers in related fields with STEM in the future with 12 questions. The adaptation of the *S survey questionnaire* was held by the researcher with the support of two specialized scientists in STEM education in an effort to increase the validity of the instrument, before the pilot test. The children completed the same questionnaire at the beginning and at the end of the activity. All the questions could be answered in a closed-type 5- grade Likert scale (absolutely disagree, disagree, neither agree nor disagree, agree, absolutely agree). The internal coherence of the questionnaire as it is described by the Cronbach' α coefficient was found 0.922 for the first part and 0.776 for the second part. The values of these coefficients show a high degree of internal coherence of the used instruments.

An evaluation questionnaire for the whole procedure and for the self-assessment of the way children worked was used also. It was consisted of twelve (12) open-ended questions.

Before the main research, a pilot test was conducted in order to test the instruments for data collection (questionnaires) with a convenient sample of a boy and a girl of the 5th grade. Of course, after the pilot test they didn't participate in the main research. In these two children were given for completion the *S survey questionnaire* and the evaluation questionnaire, to identify ambiguities, misunderstandings and difficulties in choosing a response. Based on the members' comments and process' observation, some points of the instruments were corrected. Also, defined the time required for the whole process.

The content and the characteristics of the workshops are as follows:

- 1st workshop: In the beginning, children's completed a test of knowledge and an attitudes questionnaire towards STEM sciences, which were also given (the same) at the end of the third workshop. The participants worked on an activity to be familiarized with hardware and software of LM. They constructed also, a simple robotic vehicle that they used to solve an "authentic" problem. Children worked in every group having chosen a specific role according to their preferences and skills from those which proposed in the worksheet, such as analyst, algorithm designer, technician, developer and evaluator. In addition, every group voted a representative.

- 2nd workshop: Children discussed the reasons why a solution should be found to stop the risk of air pollution. Then, they were asked about how robotics can help to solve this serious problem. Thus, they tried to create a robot that could collect things, at first without instructions, and then with instructions for those who could not make it. Finally, they programmed the robot to do exactly what they wanted after several attempts (with a reminder of the programming commands by the researcher).

- 3rd workshop: Children presented their robotic constructions in the classroom, completed the S survey questionnaire, which was the same as at the beginning of the first workshop and finally, answered the evaluation questionnaire about the advantages and drawbacks of the workshop.

6.2. Activity

The activity, which was supported by worksheets, concerns a very important problem, that of environmental pollution. Specifically, children were asked to solve an "authentic" problem (that means a problem happens in real life) and involve the rubbish collection from the waste bins of the city or from the areas surrounding the school facilities. It was emphasized that garbage areas, which are not collected, are potentially sources of contamination and a danger to the lives of citizens. So, the participants in groups (Figures 2, 3 and 4) had to use the Lego educational robotic kits and to figure out how they could find a solution to this problem. To make the lesson more experiential and practical, the problem was simulated within the classroom with empty water bottles and crinkled pieces of paper.

6.3. The Problem

Participants faced the following problem: *School cleaners strike and the waste bins and/or recycling bins of the school are overloaded. Our mission is to construct a robot that will help us to collect garbage quickly and easily, so that we do not put our health at risk by approaching and collecting them by ourselves.*

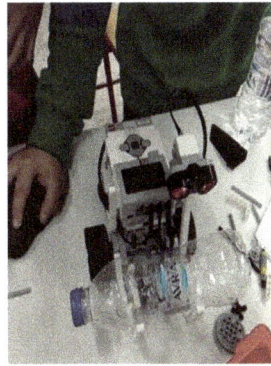  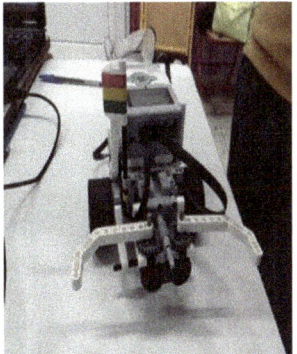

Figures 2, 3, 4: Children's robotic constructions

The five stages of solving process were:

1. problem determination,

2. searching for ideas to solve the problem (brainstorming),

3. "robo-trash" collector construction,

4. "robo-trash" collector programming,

5. running & evaluation of the program.

This activity includes: hands-on activities in the STEM education culture, working with "authentic" problem, learning through action and developing 21st century skills such as problem solving, computational thinking, work in group cooperation, etc.

7. Findings and Discussion

Pre-test & post-test data collected by the *S survey questionnaire* were analyzed with the paired samples t-test.

According to the children's answers, the smallest difference in the average scores between the two tests were found in Science questions $\bar{x} = 3.360$ for the pre-test and $\bar{x} = 3.456$ for the post-test) and the greatest difference in Engineering and Technology questions ($\bar{x} = 3.468$ for pre-test and $\bar{x} = 3.924$ for post-test). In Mathematics questions, the results were $\bar{x} = 3.68$ for pre-test and $\bar{x} = 4.00$ for post-test. Overall, it seemed the children's interaction with the educational robots was important and their attitudes become more positive. Moreover, according the questions about their careers in the future, the children responses indicated that they feel more confident and more effective in the STEM-related sciences ($\bar{x} = 3.112$ for the pre-test and $\bar{x} = 3.440$ for the post-test).

More specifically: In the field of Mathematics using the t-test was found a statistically significant difference between pre and post answers of the sample (t=-3.046; df=24; p=0.006). In the field of Science, the results did not indicate a statistical significance between pre and post answers of the children (t=-0.547; df=24; p=0.590). In the field of Engineering and Technology, the results of t-test indicated a significant statistical difference between pre and post answers of the participants (t=-3.706; df=24; p=0.01).

Finally, for pre/post-tests related to STEM's future career the results indicated a significant statistical difference between the initial and the final children's answers about their future careers they chose (t=-2.338; df=24; p=0.028).

7.1. Children's Answers to Self-Assessment Questions

According to the evaluation questionnaire, the main and meaningful results are as follows:

> At the question "What would you do to improve your effectiveness?" eight children (32%) answered that they would like to have more time, and the rest would not change anything.

> At the question "Did you work well as a group?" the majority (twenty-one) of the children (84%) answered positively, except four (16%) who faced some problems in their group.

> At the question "Were the workshops pleasant?" all the participants answered "yes".

> At the question, "What did you like most in this workshop?" all the children responded "everything" and six of them (24%) mentioned the process of robot construction and programming while seven of them (28%) programming itself as a process. Most of the children who had taken the role of the technician answered: "the construction of the robot", while the children, who took the role of the algorithm designer and the developer, answered "the programming of the robot" and it is reasonable because they chose the role by themselves according to their interests.

> At the question "Which was the most interesting thing you learned?" almost all children (96%) answered the programming of the robot and its construction.

Finally, at the question "Where could all this be useful to you?" seven children (28%) responded "to my future work" and seven more "to my daily life", which means they liked the workshop very much and think to evolve in technology.

8. Conclusions

The results obtained from the analysis of the data collected for the research question show that children's attitudes towards the STEM-related sciences have changed positively after the workshops.

At the questions about Science, there was the smallest difference in the average scores between the answers pre and post. This could be justified because the concepts of Science (e.g., friction) are difficult to understand in short term. The biggest difference in the attitudes was observed in Engineering and Technology questions. It could be attributed to the robot construction and to programming the robot. These results justify the fact that many children consider Engineering to be heavy manual work and therefore do not choose it for study and career [1]. Furthermore, as far as their willingness to pursue a career related to STEM-related sciences in the future, it appears that after the activity there were more who were interested than the beginning. It is obvious that the part of the robot's construction and programming was the one that attracted more their interest. Science and Mathematics did not have a leading role, although without them there could be no robotics. In addition, from the group's responses, we conclude that they were well aware of the dangers of mismanagement of waste and for what reason it is important to protect the environment.

Children's engagement with educational robotics in STEM activity has provided them with multiple benefits both at cognitive level and at social level. They realized the value of knowledge they acquired at school and they used it to solve a problem of everyday life. They learned to solve a problem step-by-step according to a role that they chose by themselves participation in a group. They felt the sense of responsibility and did what they did the best for their team. Their social skills were developed even more, and they worked efficiently in a group. They also developed new skills facing educational robotics. They gained experience working in the field of engineering and technology in an easy and playful way and they used pre-existing knowledge of Mathematics and Science in order to accomplish them.

In conclusion, as shown by this study and without being able to make generalizations on the results, the design of STEM courses and activities could help children to acquire all necessary skills for their future professional career but will also make them active citizens for a better society.

REFERENCES

Bowen, E., Prior, J., Lloyd, S., Thomas, S., & Newman-Ford, L. (2007). Engineering more engineers bridging the mathematics and careers advice gap. *Journal of the Higher Education Academy, 2*(1), 23–32.

Eguchi, A. (2014). Educational Robotics for Promoting 21st Century Skills. *Journal of Automation, Mobile Robotics & Intelligent Systems, 8*(1), 5–11. https://doi.org/10.14313/jamris_1-2014/1

Faber, M., Unfried, A., Wiebe, E. N., Corn, J., & Collins, T. L. (2013). *Student attitudes toward STEM: The development of upper elementary school and middle/high school student surveys* [Paper presentation]. 120th American Society for Engineering Education (ASSE) Annual Conference & Exposition, Atlanta, GA, USA. https://www.asee.org/public/conferences/20/papers/6955/view

Friday Institute for Educational Innovation (2012). *Middle/High School Student Attitudes toward STEM Survey*: Author.

Greenspan, A. (2000). The economic importance of improving math-science education. https://www.federalreserve.gov/boarddocs/testimony/2000/20000921.htm

Guzey, S. S., Harwell, M., & Moore, T. (2014). Development of an instrument to assess attitudes toward science, technology, engineering, and mathematics (STEM). School Science and Mathematics, 114(6), 271–279. https://doi.org/10.1111/ssm.12077

Johnson, C., Peters-Burton, E., & Moore, T. (2016). *STEM Road Map*. Routledge.

Komis, B. (2004). *Introduction to educational applications of ICTs*. New Technologies.

Malanchuk, O., Messersmith, E., & Eccles J.S. (2010). The ontogeny of career identities in adolescence. *New Directions for Child and Adolescent Development, 130*, 97–110. https://doi.org/10.1002/cd.284

National Academy of Engineering (2008). *Grand challenges for engineering*. The National Academies Press.

Psycharis, S. (2016). Inquiry Based- Computational Experiment, Acquisition of Threshold Concepts and Argumentation in Science and Mathematics Education. *Educational Technology & Society Journal, 19*(3), 282–293.

Riberio, C., Coutinho, C., Costa M.F.M., & Rocha, M. (2008). A study of educational robots in elementary schools. In M. F. P. Costa, & J. B.V. Dorrio (Eds.), *Selected Papers on Hands-On Science* (pp. 580–595). Hands-On Science Network.

Stergiopoulou, M. Karatrantou, A., Panagiotakopoulos, C. (2016). Educational Robotics and STEM Education in Primary Education: Apilot study using the H&S Electronic Systems Platform. In D. Alimisis, M. Moro, & E. Menegatti (Eds.), *Educational Robotics in the Makers Era* (pp. 88–103). Springer International Publishing.

Suprapto, N. (2016). Students' Attitudes towards STEM Education: Voices from Indonesian Junior High Schools. *Journal of Turkish Science Education, 13*, 75–87.

Tseng, K. H., Chang, C. C., Lou, S. J., & Chen, W. P. (2013). Attitudes towards science, technology, engineering and mathematics (STEM) in a project-based learning (PjBL) environment. *International Journal of Technology and Design Education, 23*(1), 87–102. https://doi.org/10.1007/s10798-011-9160-x

Tyler-Wood, T., Knezek, G., & Christensen, R. (2010). Instruments for assessing interest in STEM content and careers. *Journal of Technology and Teacher Education, 18*(2), 345–368.

United States Department of Commerce (2012). *The Competitiveness and Innovative Capacity of the United States*. USA Department of Commerce.

Wing, J.M. (2006). Computational thinking. *Communications of the ACM, 49*, 33–35. https://doi.org/10.1145/1118178.1118215.

Chapter 15

From STEM to STEAM and to STREAM Enabled Through Meaningful Critical Reflective Learning

Vassilios Makrakis[1]

[1]Professor & UNESCO Chairholder in ICT in Education for Sustainable Development, Department of Primary Education, University of Crete, makrakis@edc.uoc.gr

Abstract: *Integrating Science, Technology, Engineering and Mathematics (STEM) subjects can be a rewarding process in terms of promoting meaningful learning that is engaging learners among other things in problem-solving, critical thinking and in building real-world connections. However, such a vision is still a quest and STEM or STEAM in its new version has been an area of controversy in terms of outcomes. The question is; how can the content and processes of the four individual subjects become integrated learning areas? How can the integrity of each of these subjects areas be maintained and yet be integrated in a meaningful way? In this paper, I argue that critical reflection being a driving force towards making STEAM (including Arts) learning more integrated, meaningful and engaging for the students. In this sense, STREAM, integrating R (Reflective learning) is justified and discussed.*

Keywords: *STEM education, STEAM, STREAM, critical reflection, meaningful learning*

1. From STEM to STEAM (STEM + The Arts) and Meaningful Learning

STEM is an acronym for Science, Technology, Engineering and Mathematics and its affiliated subjects. Integrating STEM helps students develop relevant knowledge, concepts and skills as well as connect relevant skills in real world applications (Brophy et. al, 2008). However, the continued separation of the STEM subjects, in terms of how, when and where they are taught continues to occur in all education levels for a number of administrative and organisational reasons (Herschbach, 2011). Hom (2014) states that, although the United States has historically been a leader in the STEM fields, fewer students have been focusing on the STEM topics until recently. A small percentage (16%) of high school students in the US are interested in a STEM career and have proven a proficiency in mathematics, despite the huge investments in promoting STEM education. The same source reveals that US consistently ranks poorly against its global classmates in STEM subjects, placing 25th in math and 17th in science out of 31 countries ranked by the Organization for Economic Cooperation and Development (ibid.). STEM Education spans across all education levels, from primary to secondary and tertiary education. At the primary and secondary level, STEM education focuses on using math, technology and science to solve real-world problems by applying project/problem-based teaching and learning methodologies. The international results, however, are very disappointed not only in US but also in many countries (OECD, 2018). One of the main reasons for the continued separation of the STEM subjects comes from the fact that the teachers (especially in high school and universities) come from different discipline backgrounds and each values their domain of knowledge as a separate area of learning with its own history and curriculum practices (Herschbach, 2011).

It seems that STEM education is perpetuating the domination of the so-called "hard" science and in doing so is promoting an unsustainable economic growth model. The recent incorporation of Arts into STEM, turning it to STEAM (STEM + the Arts), maybe seen as a reaction to the lack of its integrative practices. The Arts covering a wide range of related subjects representing humanities adds a new dimension. Although putting Arts in the other STEM subjects does not necessarily imply a more integrative approach, it creates optimism for more holistic approaches to teaching and learning.

Gunn (2017) in answering the question: Why the A in STEAM is just as important as every other letter, argues that every engineer besides skills in STEM also needs design- thinking, creativity, communication, and artistic skills to bring those innovations to fruition. Sochacka et al. (2013) highlight potential difficulties associated with current understandings of STEAM education, arguing that insufficient attention has been given to what STEM disciplines might contribute to the arts. In general, STEAM is built on the premise that there is space and need for cross-fertilization among its disciplines and any interdisciplinary and cross-disciplinary initiative offers reciprocal benefits to the learners and disciplines involved. It is my conviction that reasons why STEM with Arts is leading to a superficial integration lies in the following three factors: 1) the narrow approach of viewing Arts; 2) the orientation of investment in STEM to staff recruitment and less on teacher quality and 3) the lack of a sound and broader pedagogical approach to teaching STEAM.

At the higher education level, various international initiatives have attempted to bridge the distorted distinction between "hard" and "soft" sciences. The European RUCAS (Reorient University Curricula to Address Sustainability)-Tempus project implemented in six universities in Egypt, Lebanon and Jordan leading to the training of 100 academic staff from six academic disciplines, including STEAM to revise more than 200 courses to infuse sustainability issues (Makrakis & Kostoulas-Makrakis, 2013a, 2013b, 2016). Building on RUCAS, another European- funded Tempus project, CLIMASP (Climate Change and Sustainability Policy) attempted to transform current practices that prevent inter/cross-disciplinary collaboration among academic staff by developing the CLIMASP interdisciplinary minor program integrated into a regular undergraduate degree in disciplines like education sciences, technical sciences, engineering, mathematics, sciences, economics/business and social sciences (Makrakis & Kostoulas-Makrakis, 2015). Students from STEAM can take courses from three concentration areas, namely, (1) Climate Change, Environment and Society; (2) Climate Change, Economics and Public Policy; and (3) Climate Change, Science and Technology. Building on CLIMASP, another European funded Tempus program, CCSAFS aims to develop an inter/multidisciplinary MSc programme in Climate Change, Agricultural Development and Food Security in Egypt and Jordan.

Experiences gained through the RUCAS, CLIMASP and CCSAFS inter/cross- disciplinary projects show that STEAM become meaningful when a broader pedagogy engaging critical reflection and critical discourse, whereby the learner is turning able to think and act "out of the box", is the driver making integrative learning more meaningful. The key characteristics of meaningful learning are the following:

- **Reflective** – learning that involves students in a process to self-critical assessment of their learning experiences, identify areas that require improvement and proceed in constructing new knowledge that makes a difference.

- **Active** – learning that involves students in a process that requires them to play an active role to construct knowledge and understanding.

- **Experiential** – learning that involves students in a process whereby they reflect on, learn from, develop new knowledge, and take new action based on experience.

- **Constructive** – learning that involves students in a process of experiencing things and reflecting on those experiences to constructing understanding, knowledge and meaning of the world.

- **Transformative** – learning that involves students in a critical self-reflecting process of deconstructing, constructing and reconstructing themselves and social realities.

- **Collaborative** – learning that involves students to construct meaning and knowledge collectively and collaboratively.

- **Dialogical** – learning that involves two or more students in a process of structured, purposeful egalitarian dialogue with a shared goal of raising their critical consciousness about an issue that concerns them.

- **Political** – learning that involves students in a process of reflecting and acting on themselves and the world in order to transform it.

- **Ethical** – learning that involves students in a process of reflection and action towards the common good driven by moral principles and values.

- **Authentic** – learning that involves students in a process that allows them to explore, discuss, and meaningfully construct concepts and relationships in contexts that involve real-life problems.

- **Problem-posing** – learning that involves students in a process of identifying, coding and decoding real-life issues through critical discursive reflection and acting to change realities.

- **Subversive** – learning that involves students in a process of raising critical questions about teaching, learning and curriculum and their role and underlying assumptions about these processes.

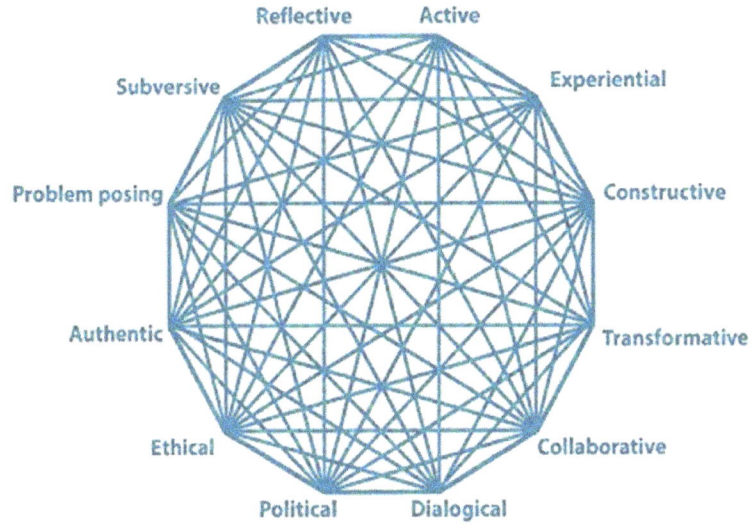

2. The Missing R from STEAM: Shifting to STREAM (STEAM + Reflective Learning)

Looking into the meaningful learning key components (Figure 1) defined previously it becomes obvious that critical reflective learning is adding an important dimension in each component. Thus, turning STEAM learning meaningful, it is necessary to integrating on STEAM, the R (reflective learning), and turning STEAM into STREAM. Why the R in STEAM is very critical? In answering this question, I will take a personal experiential approach.

One of the questions that has driven my academic teaching is the following: How does disciplines or subjects such as sciences, technology, engineering, arts and mathematics look and act when viewed in/through the lenses of sustainability justice? The concept of sustainability justice I have developed and applied widely in my teaching reflects the four pillars of sustainable development: environment, society, economy and culture from a justice perspective. In particular, sustainability justice is perceived as a process, and not an outcome, which: 1) seeks fair (re)distribution of resources, opportunities, and responsibilities; 2) challenges the roots of oppression and injustice; 3) empowers all people to raise their voice, needs and right; and 4) constructs knowledge, empathy, compassion, social solidarity and action competences (Makrakis, 2017). Infusing sustainability justice in teaching STEAM makes their integration and meaningfulness possible.

In my attempt to infuse sustainability justice in the RUCAS, CLIMASP and CCSAFS curricula, instructors in the fields of sciences, technology, engineering and mathematics could not grasp this possibility and even could not understand its necessity. Positivistic and instrumentalist assumptions in separating knowledge and values have led many to view these subjects and arts (including sustainability justice) as being positioned on opposing ends of a spectrum describing the first as "hard sciences" (quantitative reasoning) and the second as "soft sciences" (qualitative reasoning) and, thus, incompatible for integration. It is this false dichotomy and the illusion of academic segregation that obstacles the integration of STEAM subjects and above all their relevance to social reality. Bond and Chernoff (2015) argue that, "social justice provides engaging, empowering, and authentic contexts for projects in which mathematics skill sets can come alive and transcend the traditional limited and abstract operations that have isolated and discouraged too many students for too long" (p. 29). They also state that, "social justice can elicit intrinsic motivation in students, which inspires growth far beyond the basic traditionally requisite mathematical skill set and that the two strands are best contextualized authentically within the local community (ibid.). In the radical math website (http://www.radicalmath.org/) there are many free lesson examples that show how to merge mathematics and sciences with sustainability justice dimensions. The link between mathematics and sciences with social justice has been well documented and justified (Garii & Appova, 2013; Gutstein, 2006; Gutstein & Peterson, 2013; Skovsmose, 1994).

Posing sustainability justice questions implies, first, that my role in teaching and learning cannot be perceived as a neutral action and second, my teaching is seen as a political praxis guided by a moral disposition (an ethical framework) or *phronesis* in Aristotelian terms, that is, to act truly and rightly having in mind the common good (Aristotle, ca. 350 B.C.E./2004, p. 209). My role and interests are also explicit to my students, since "knowing ourselves" and "others knowing yourselves" prevents the indoctrination and imposition of ones values uncritically.

I envision my students functioning as agents of change by helping them in developing their critical consciousness. It is not my job to provide answers to the real-life problems they face, but to help them achieve a form of critical reflective thinking about the situation through moving between theory, action and reflection on action. Freire (1974) called it conscientization, that is, to understand that the world or society is not fixed and is potentially open to transformation by active citizens. I encourage my students to get involved in an inner and outer dialogue undertaken jointly with others in order to understand "who they are", recognize, challenge and even change themselves and social reality. Learning to me is thus a continuous process, directed at enhancing the learners' capacity to act in the world and change it. Respectively, teaching to me is an ethical and political praxis. Praxis enables me to reflect on my roles and actions by making the political more pedagogical and the latter more political. Guided by Freire's pedagogy and Mezirow's (1990) transformational theories, I also see praxis as the vehicle through which we evaluate our own worldviews, habits of mind and the meaning perspectives with which we have made sense out of our encounters with the world, others, and ourselves. This crucially important personal learning dynamic is analogous to the process of paradigm shift that Thomas Kuhn characterized as the way revolutions occur in science. In teaching, praxis also implies that abstract theorising is only useful so long as it informs concrete action and consciousness-raising through reflection and action. Through the iterative processes of consciousness, praxis and critical reflection, I become aware of my teaching actions and their ethical implications. Praxis without critical consciousness and critical reflection replicates what makes human life unsustainable and at the same, it is reinforcing and perpetuating the existing social, environmental, economic and cultural injustices. It is through praxis, dialogue and reflection that meaningful learning can be better achieved in and outside the classroom.

It becomes, thus, clear that teaching critically and reflectively is an act of resistance. It refers to a way to resist domestication and disempowerment, a negative process of socializing learners into conformity to established norms and values, preventing them from learning to clarify, deconstruct and reconstruct what they want to be, to know, and do. Transforming from a domesticated situation into an empowering experience becomes possible through dialogue that leads to deep understanding of the situation, new knowledge construction, meaning making and praxis in the form of change. I teach my students that it is not enough for people to come together in dialogue in order to gain knowledge of their and others' social reality. They must act together upon their environment in order critically to reflect upon their reality and so to transform it through further action and critical reflection. In this context, in my attempt to embed sustainability justice issues in STEAM curricula, some of the problem-posing questions are:

- What aspects of ourselves can we see through techno-sciences and mathematics?

- How can we look into the future through techno-sciences and mathematics?

- Do sciences, technology and mathematics have social meaning and constitute a critical factor for social change?

- How can, for example, new technology helps: to integrate science and mathematics with other subjects; to promote more cooperative learning, to encourage the transfer of science and mathematics process skills to everyday life, and to improve student attitudes towards science and mathematics?

- How do science and math teachers could help learners make meaningful social meaning out of science and math problems?

Posing such issues and questions turns teaching as a subversive activity by putting forward ideas about teaching, learning and curriculum that challenge and eliminate product-oriented STEAM curricula and textbooks, teacher-centered teaching and transmissive learning practices.

Critical reflection is thus ethically informed and those involved are committed to transformative action. Moving from theory to praxis helps to challenge the technical cognitive interest driven by instrumental rationality that has pervaded STEM education in the 20th century. As defined by Habermas (1972), instrumental learning and technical knowledge interest is best understood as learning to control and manipulate learners or the environment. This is in contrast to practical knowledge interest associated with communicative or transactive learning, which is concerned with understanding and critical consciousness. Understanding by itself cannot change reality. There is need to merge communicative learning to emancipatory learning in order to make a positive difference.

For teachers functioning as reflective practitioners and problem-posing educators, curriculum is seen as, a process and praxis, rather than a product to be transmitted. In the context, teachers have to see themselves functioning as facilitators and mentors, as resources and as curriculum developers (Makrakis, 2006) as well as transformative intellectuals (Giroux & Mclaren, 1996). Each of these roles is associated with specific activities. Teachers as "facilitators and mentors" will guide and facilitate learners' critical and creative thinking in a collaborative learning environment enabled by new technology. Teachers as "resources" will have to develop learners' capacities for active citizenship and to contribute to their fellow teachers' professional development enabled by new technology. As "curriculum developers", teachers critically assess school knowledge, reorder and enrich curriculum according to the principles of new pedagogy enabled by new technology. Teachers as "transformative intellectuals" are involved in developing a discourse that unites the language of critique with the language of possibility (Giroux, 1988). Teachers functioning as "transformative intellectuals" are giving students an active voice, making the political more pedagogical and the latter more political (Giroux, n.d.). In other words, teachers are able and committed to function as change agents of reorienting what they teach and how they teach for promoting a key goal for 21st Century education that is, learning to transform oneself and society.

3. Concluding Remarks

Today developing an understanding on issues related to sustainability has gained a renewed concern among STEAM educators. It is often debated that techno-science (technology and science) and mathematics, although they have generated enormous positive developments across all societal spheres, they have also brought worldwide problems, some of which are irreversible. My experiences with Greek and other countries' school curricula and teaching methodologies show that are largely driven by instrumental rationality and technical interest in knowledge, which does little to develop human self-realization and critical discourse. Science and especially mathematics curricula seem to be treated from a surface learning perspective and largely decontextualized without social meaning. Thus, science and mathematics are disassociated from social reality and do not serve as vehicles for better local/global understanding of the real-life issues that students encounter in everyday life.

A study undertaken by Makrakis & Kostoulas-Makrakis (2005) has posed the above questions and revealed that the Greek primary school curriculum in science and mathematics was decontextualized and lacked a connection to the real-life issues that learners encounter outside school. The results of that study showed that beyond science and mathematical literacy, almost all mathematics and science lessons consisted of content that was divorced from any significant understanding of social processes and realities faced outside the school walls. Grasping only STEAM literacy without critical reflective learning makes STEAM education meaningless. Accordingly, school does not engage learners in experiences that are relevant to life and their development as socially responsible and critically thinking citizens.

Looking into the current Greek Science and Mathematics textbooks, the content titles are significantly different from those textbooks researched 15 years ago in terms of opportunities to embed sustainability justice issues. For example, in the 6th Grade Science (Physics) curriculum out of the nine chapters included in the learner's textbook, three chapters can potentially have a direct link to education for sustainability, namely: Energy (pp. 24–31); Digestive System (pp. 32–39); Heat (pp. 40-55) and Electricity (pp. 56–70). The unit dealing with Energy relates energy with everyday life as it proclaims, including the chapters of energy depositories or sources of energy as well as food and energy. In the Mathematics textbook, compared to the Sciences textbook, surprisingly, most of the Chapters (55 in total) in the nine Units could have a direct link with the four pillars of sustainable development (environment, society, economy and culture). For example, in Unit 1 includes Chapter 3 "The Greeks of Diaspora" (pp.16–17), in Unit 2, Chapter 11 "In the restaurant" (pp. 34–35), in Unit 3, Chapter 17 "Elections in the class" (pp. 48–49) and Chapter 19 "Selecting the most economic packaging" (pp. 52–53). In Unit 4, Chapter 23 "Choosing what we eat" (pp. 64–65) and Chapter 24 "Careta-Careta" (pp. 66–67). Besides that, the 6th Grade Mathematics textbook includes two Chapters dealing with Technology, namely Chapter 7, in Unit 2 "In the Informatics Lab" (pp. 26–27) and Chapter 48 in Unit 8 "Games in the computer" (pp. 120–121).

Looking into these units and chapters one would expect curriculum connections of sciences and mathematics literacy and skills with the real-life issues denoted explicitly or implicitly in the titles of these units. However, there is not any real and deep learning intention to make cross-curricular connections contextualised with sustainability issues. There is absence of meaningful learning and sustainability justice methodology that would turn these subjects approached in a horizontal, integrative and holistic way. They reflect an instrumental rationality and linearity in knowledge representation, following a hierarchical structure. Despite that curriculum guidelines are sometimes referring to higher levels of Bloom's Taxonomy in the objectives set, what one can encounter in the content is these textbooks is a focus at the competence level of describing, explaining and demonstrating, with less opportunities for meaningful learning.

As pointed earlier, STEAM school curricula are not sustainability justice-free or neutral subjects. Integrating the R (critical reflective learning) into STEAM and turning it to STREAM can provide meaningful and challenging contexts for developing a wide range of sciences, ICT skills, engineering, and mathematics literacy and skills in an integrative and holistic way. Lessons that address issues of equity, gender, cultural diversity and intercultural understanding, health, HIV/AIDS, governance, natural resources, climate change, rural development, sustainable urbanisation, poverty reduction, corporate responsibility and accountability can effectively be woven into STEAM classes in culturally responsive and motivating creative ways (Makrakis, 2011a, 2011b, 2012, 2014; Makrakis & Kostoulas-Makrakis, 2017). STEAM teachers should act as reflective practitioners providing learners opportunities for critical reflective learning while studying STEAM. Thus, the introduction of STREAM acronym could give space for further discussion aiming to make learning more meaningful and culturally relevant for students.

Acknowledgments

This work has been developed within the framework of the CCSAFS (Development of an interdisciplinary MSc programme in climate change, sustainable agriculture and food security) project that has been funded from the European Commission (No. 573881-EPP-1-2016-1-EL-EPPKA2-CBHE-JP (2016-3770-001-001). The content of the paper reflects the views of the authors, and the Commission cannot be held responsible for any use which may be made of the information contained therein.

REFERENCES

Aristotle (2004). *The Nicomachean ethics* (H. Tredennick, Ed. & J. A. K. Thomson, Trans.). Penguin. (Original work published ca. 350 B.C.E.)

Bond, G., & Chernoff, E. (2015). Mathematics and Social Justice: A Symbiotic Pedagogy. *Journal of Urban Mathematics Education, 8*(1), 24–30.

Brophy, S., Klein, S., Portsmore, M., & Rogers, C. (2008). Advancing engineering education in P-12 classrooms. *Journal of Engineering Education. 97*(3), 369–387. https://doi.org/10.1002/j.2168-9830.2008.tb00985.x

Freire, P. (1974). *Education for critical consciousness*. Continuum.

Garii, B., & Appova, A. (2013). Crossing the great divide: Teacher candidates, mathematics, and social justice. *Teaching and Teacher Education, 34*, 198–213. https://doi.org/10.1016/j.tate.2012.07.004

Giroux, H. (n.d.). *Teachers as transformatory intellectuals* [Symposium]. Symposium on understanding Quality Education: Conference on Re-envisioning Quality Education. http://www.afed.itacec.org/document/henry_giroux_2_ok.pdf

Giroux, H. (1988). *Teachers as intellectuals*. Bergin & Garvey.

Giroux, H. (2007, February 18). Μάθημα αμφισβήτησης της εξουσίας: συνέντευξη στον Χρόνη Πολυχρονίου [Lesson of questioning power: Interview to Chronis Polychroniou]. *Κυριακάτικη*. https://www.aua.gr/gr/synd/eedip/Nea/2007/Pan_eleyth_07-02-18_Polyhroniou.pdf

Giroux, H. A. & Mclaren, P. (1996). Teacher education and the politics of engagement: The case for democratic schooling. In P. Leistyna, A. Woodrum, & S.A. Sherblom (Eds.), *Breaking free: The transformative power of critical pedagogy* (pp. 301–331). Harvard Educational Review.

Gunn, J. (2017, November 8). Why the A in STEAM is just as important as every other letter? *Room 241*. https://education.cu-portland.edu/blog/leaders-link/importance-of-arts-in-steam-education/

Gutstein, E. (2006). *Reading and writing the world with mathematics: Toward a pedagogy for social justice*. Routledge.

Gutstein, E., & Peterson, B. (2013). *Rethinking mathematics: Teaching social justice by the numbers* (2nd ed.). Rethinking Schools.

Habermas J. (1972). *Knowledge and Human Interests* (2nd ed.). Heinemann

Herschbach, D. R. (2011). The STEM initiative: Constraints and challenges. *Journal of Stem Teacher Education*, 48(1), 96–122.

Hom, E. (2014). What is STEM Education? *Live Science*. https://www.livescience.com/43296-what-is-stem-education.html

Makrakis, V. (2006). *Preparing United Arab Emirates Teachers for Building a Sustainable Society*. E-Media Publications: University of Crete.

Makrakis, V. (2011a). Strategies for change towards sustainability in tertiary education supported by ICT. In UNESCO Institute for Information Technologies in Education (Ed.), *ICT in Teacher Education: Policy, Open Educational Resources and Partnership. Proceedings of International Conference IITE-2010* (pp.152–166).

Makrakis, V. (2011b). ICT-enabled education for sustainable development: Merging theory with praxis. In M. Youssef & S. Aziz Anwar (Eds.), *Proceedings of the 4th Annual Conference on e-Learning Excellence in the Middle East 2011 In Search of New Paradigms for re-Engineering Education* (pp. 410–419). Hamdan Bin Mohammed e-University.

Makrakis, V. (2012). Reorient Teacher Education to address Sustainable Development Issues through the WikiQuESD. In A. Jimoyiannis (Ed.), *Research on e-Learning and ICT in Education* (pp. 83–94). Springer.

Makrakis, V. (2014a). ICTs as transformative enabling tools in education. In R. Huang, Kinshuk & J. Price (Eds.), *ICT in education in global context* (pp. 101–119). Springer Verlag.

Makrakis, V. (2014b). Transforming university curricula towards sustainability: A Euro-Mediterranean initiative. In K. Tomas & H. Muga (Eds.), *Handbook of Research on Pedagogical Innovations for Sustainable Development* (pp. 619–640). IGI Global.

Makrakis V. (2017). Unlocking the potentiality and actuality of ICTs in developing sustainable–justice curricula and society. *Knowledge Cultures*, 5(2), 103–122. https://doi.org/10.22381/KC5220177

Makrakis, V., & Kostoulas-Makrakis, N. (2005). Techno-sciences and Mathematics: Vehicles for a sustainable future and global understanding. In P. G. Michaelides & A. Margetousaki (Eds.), *Proceedings of 2nd International Conference on Hands on Science. HSci 2005* (pp. 103–108). E-Media: University of Crete.

Makrakis, V., & Kostoulas-Makrakis, N. (2013a). Sustainability in higher education: A comparative study between European Union and Middle Eastern universities. *International Journal of Sustainable Human Development*, 1(1), 31–38.

Makrakis, V., & Kostoulas-Makrakis, N. (2013b). A methodology for reorienting university curricula to address sustainability: The RUCAS-Tempus project initiative. In S. Caeiro, W. Leal Filho, C. Jabbour, & U. Azeiteiro (Eds.), *Sustainability assessment tools in higher education institutions* (pp. 23–44). Springer International Publishing Switzerland. https://doi.org/10.1007/978-3-319-02375-5

Makrakis, V., & Kostoulas-Makrakis, N. (2014). An instructional-learning model applying problem-based learning enabled by ICTs. In A. Anastasiadis, N. Zaranis, V. Oikonomidis, & M. Kalogiannakis (Eds.), *Proceedings of the 9th Panhellenic Conference on ICTs in Education* (pp. 921–933). University of Crete.

Makrakis, V., & Kostoulas-Makrakis, N. (2015). A strategic framework for developing interdisciplinary minors on climate change and sustainability policy: The CLIMASP-Tempus example. In W. Leal Filho, W. Azeiteiro, S. Caeiro, & F. Alves (Eds.), Integrating sustainability thinking in science and engineering curricula (p.103–114). World Sustainability Series. Springer International Publishing Switzerland. https://doi.org/10.1007/978-3-319-09474-8

Makrakis, V., & Kostoulas-Makrakis, N. (2016). Bridging the qualitative– quantitative divide: Experiences from conducting a mixed methods evaluation in the RUCAS programme. *Evaluation and Program Planning, 54,* 144–51. https://doi.org/10.1016/j.evalprogplan.2015.07.008

Makrakis, V., & Kostoulas-Makrakis, N. (2017). An instructional-learning model applying problem-based learning enabled by ICTs. In P. Anastasiades & N. Zaranis (Eds.), *Research on e-Learning and ICT in Education* (pp. 3–16). Springer.

Mezirow, J. (1990). *Fostering critical reflection in adulthood: a guide to transformative and emancipatory learning.* Jossey-Bass.

OECD (2018). *PISA 2015: Results in Focus.* Paris: OECD. https://www.oecd.org/pisa/pisa-2015-results-in-focus.pdf

Skovsmose, O. (1994). *Towards a philosophy of critical mathematics education.* Kluwer.

Sochacka, N., Woodall Guyotte, K., Walther, J., & Kellam N. N. (2013). *Faculty reflections on a STEAM-inspired interdisciplinary Studio course* [Paper presentation]. 120th American Society for Engineering Education (ASSE) Annual Conference & Exposition, Atlanta, GA, USA. https://www.asee.org/file_server/papers/attachment/file/0003/4055/6555.pdf

Chapter 16

Bridging STEM (Science, Technology, Engineering & Mathematics) Education with Education for Sustainability

Nelly Kostoulas-Makrakis[1]

[1]*Associate Professor, Department of Primary Education, University of Crete, nkostoula@edc.uoc.gr*

Abstract: *STEM, otherwise known as Science, Technology, Engineering and Mathematics, accounts for most of the skills and knowledge needed in the 21st century. As such, STEM, although originated two decades ago is now making a new impact into education at all levels. My argument is that STEM, and hence STEM Education discourse has focused more on an instrumental perspective situated in an economic growth ideology that has driven current unsustainable development. In this paper, I present a critique of the mainstream STEM Education from an Education for Sustainability (EfS) perspective, which argues that STEM needs to shift towards a new paradigm encompassing the principles of EfS. In this way, I outline the principles of EfS and explore how EfS and STEM Education might be brought together, providing examples of European Commission funded projects that integrate STEM and EfS.*

Keywords: *STEM, RUCAS, CLIMASP, CCSAFS, Education for Sustainability*

1. Mainstream STEM Education and Its Contradiction with Efs

In the last two decades, there are numerous policies to embed the goals, knowledge, skills and values, which support sustainable development across all disciplines, including the disciplines of Science, Technology, Engineering and Mathematics (STEM). On 25 September 2015, the UN General Assembly adopted the 2030 Agenda for Sustainable Development. At the core of the UN 2030 Agenda are the following 17 Sustainable Development Goals (SDGs) cited by UNESCO [2017, p.6]. The aim of the 17 SDGs is to secure a sustainable, peaceful, prosperous and equitable life on earth for everyone now and in the future. The goals cover global challenges that are crucial for the survival of humanity.

1. No Poverty
2. Zero Hunger
3. Good Health and Well-Being
4. Quality Education
5. Gender Equality
6. Clean Water and Sanitation
7. Affordable and Clean Energy
8. Decent Work and Economic Growth
9. Industry, Innovation and Infrastructure
10. Reduced Inequalities
11. Sustainable Cities and Communities
12. Responsible Consumption and Production
13. Climate Action
14. Life below Water
15. Life on Land
16. Peace, Justice and Strong Institutions
17. Partnerships for the Goals

Sterling (2004, p.50) pointed out that sustainability "is not just another issue to be added to an overcrowded curriculum, but a gateway to a different view of curriculum, of pedagogy, of organisational change, of policy and particularly of ethos". Embedding education for sustainable development (ESD) within the curricula for all undergraduate students, especially in teacher education becomes of critical importance to quality education. From a teaching methodological perspective, ESD pedagogy aligns with interdisciplinary and transdisciplinary educational methods and approaches: a) to develop an ethic for lifelong learning; b) to foster respect for human needs that are compatible with sustainable use of natural resources and the needs of the planet; and c) to nurture a sense of global solidarity (UNESCO, 2010). According to Wals (2012), "at least four lenses of ESD can be distinguished:

- **An integrative lens:** taking on a holistic perspective that allows for integrating multiple aspects of sustainability (e.g., ecological, environmental, economic and sociocultural; local, regional and global; past, present and future; human and non-human).

- **A critical lens:** questioning predominant and/or taken-for-granted patterns and routines that are or may turn out to be unsustainable (e.g., the idea of continuous economic growth, dependency on consumerism and associated lifestyles).

- **A transformative lens:** moving beyond awareness to incorporate real change and transformation through empowerment and capacity-building that may lead to or allow for more sustainable lifestyles, values, communities and businesses.

- **A contextual lens:** recognizing that there is no one way of living, valuing and doing business that is most sustainable everywhere and always and that although we can learn from each other, places and people are different and times will change. Therefore, sustainability needs to be recalibrated as realities and times change".

Smith and Watson (2016) argue that the promotion of K-12 STEM education is directed towards the neoliberal project of producing applied specialists to enhance economic growth and competitiveness through technological solutions. All these imply revising teaching content to respond to global and local challenges through the development of certain skills. In the second half of the 20th century, much of the discussion on skills needed centered on the 3Rs - reading, writing and arithmetic. In the last decade, there is a shift to what has been termed as the 4Cs - critical thinking & problem solving, communication, collaboration and team building, creativity and innovation (American Management Association, 2010; AT21CS, 2012; Partnership for 21st Century Learning, 2011). Recently Makrakis (2017) considered the 4Cs model inadequate and expanded it to the 10Cs model, taking into consideration the challenges posed by the sustainability crisis and the expansion and potential of ICTs as enabling tools for transformative learning. As pointed by Makrakis (2017), although there is some overlap among the 10Cs, each one has its own role in teaching and learning for problem solving. For example, critical thinking and problem solving refers to the ability to make decisions, solve problems and take appropriate action, using learning processes such as conceptualizing, applying, analyzing, synthesizing and/or evaluating information gathered by multiple means. These skills are largely related to supporting sustainable behavior, civic engagement, as well as viable employment and a better quality of life. Besides the 10C skills that support Education for Sustainability (EfS), there are also a number of skills suitable to support EfS experiences, processes and practices such as, conflict resolution, creative, imaginative, real-world problem-solving, negotiation and participatory action research.

Skills alone cannot make a difference unless supplemented by certain values. Murray and Murray (2007) stress that education must go beyond knowledge and skills to include the encouragement of values that support sustainable development issues. Learning to clarify one's own values, rather than imposing values without reflection is a very critical domain of learning goals and teaching methodology.

In assessing the link between STEM and EfS, the following questions should be asked: Do your school's STEM courses frequently use sustainability themes and EfS principles and methodologies as a context for learning? Does your school curriculum make connections between school knowledge and real-life problems found out the school? If you are a teacher, do you involve your students in problem-posing questions, developing and using models, planning and carrying out investigations, analyzing and interpreting data, constructing explanations, and engaging in argument from evidence when exploring environmental and sustainability issues?

Through my experience in teaching pre-service teachers in courses related to teaching methodology and supervising future teachers in school teaching practicum, I have realized that, although currently used Greek school textbooks integrate some principles of the above SDGs, the way they are embedded and the content presented is more ascribing to surface than deep-learning. The concept of a deep approach to learning originated in the work of Marton and Säljö (1976, 1984). Deep level learning is associated with those learners who attempt to relate ideas together to understand underpinning theory and concepts, and to make meaning out of material under consideration (Dolmans et al., 2016). It is also associated with those learners who are able to understand authors' and/or lecturers' words enough to give a meaning using their own words (Marton & Säljö, 1984). According to Warburton (2003), deep learning is a key strategy by which students extract meaning and understanding from course materials and experiences. According to the Bologna declaration, successful learning and studying in higher education should involve students in deep learning (Asikainen, 2014). STEM can be powerful tools in the movement towards sustainability through proper contextualization and deep learning. High-quality STEM teaching and learning should be relevant and meaningful to students' lives, giving particular importance to the interdisciplinary thinking and holistic insight, making curriculum connections, applying experiential, constructivist and transformative learning as well as critical reflective practice.

2. Interventions to STEM through RUCAS, CLIMASP and CCSAFS Projects

RUCAS (Reorient University Curricula to Address Sustainability) was a Tempus funded project coordinated by the University of Crete implemented in the period 2010- 13 aimed to: 1) Support the development of ESD in the Higher Education sector in Egypt, Jordan and Lebanon. 2) Build capacity amongst university staff to embed ESD in curricula and pedagogy and 3) Review and revise undergraduate curricula to address ESD in line with the Bologna and Lisbon processes. University teaching staff from three MENA region countries (Egypt, Jordan and Lebanon) along with Universities from France, Italy and Sweden critically reflected on the content of their courses and teaching methods to see what gaps and what emphases were missing in relation to sustainability. In particular, the tasks involved in the process of course revision enabled university instructors to identify objectives for ESD that suit their subject area and the content that is missing, then proceed to matching both objectives and content as well as what is suitable to ESD teaching/learning methods (e.g., values clarification, problem-based learning, critical reflection).

The ESD themes infused in STEM as well as in other subject such as economics/business and social sciences ranged widely, and the key strategic themes included climate change; energy use and management; sustainable urbanization; natural resources (water security, deforestation, sustainable agriculture, biodiversity); child labour, sustainable tourism, fair trade, social justice; indigenous knowledge; sustainable production/consumption. The tasks involved in the process of creating the revised course syllabus focused on identifying key learning goals/objectives and outcomes, formulating appropriate feedback and assessment procedures, and selecting and developing suitable teaching/learning activities, mostly student-led.

The RUCAS Toolkit (http://www.rucastoolkit.eu/) a collection of tools and resources structured within eight modules, and an online community of practice and related resources can be used to assist university instructors and other staff in reorienting university curricula to address sustainability. More than 4,000 students were involved in the monitoring and evaluation activities. Students that participated in practicum placements totalled 1,861 during the autumn semester 2012-13. The general trend was that almost all the themes of the practicum assignments were contextualized in the local environment (Makrakis & Kostoulas-Makrakis, 2013a, 2013b, 2016).

The CLIMASP (Climate Change and Sustainability Policy) was a Tempus funded project built on the RUCAS project and coordinated by the University of Crete during a three-year period (2013-2016). The wider CLIMASP objectives were to transform current unsustainable practices that prevent interdisciplinary collaboration and promote sustainable leadership in the partners' countries universities in Egypt, Jordan and Lebanon) and its graduates. The specific objectives were to:

- Develop capacity-building programmes to train university teaching staff and key administrators for interdisciplinary collaboration and building partnerships with local/national/regional partners.

- Involve university staff and other key stakeholders (e.g., students, professionals) in the development of an undergraduate interdisciplinary programme on CLIMASP in each partner country University.

- Integrate and implement the CLIMASP programme as an integral part to existing undergraduate academic degrees in STEM disciplines as well as education sciences, economics/business sciences and social sciences.

In particular, the minor in Climate Change and Sustainability Policy that has been developed and implemented offers students a unique, inter/multidisciplinary understanding of climate change. Choosing a minor combined with a major degree it gives students the opportunity to pursue a second area of interest that can open significant career opportunities, besides building knowledge and skills in an area that fulfils their personal interests. The interdisciplinary CLIMASP courses are integrated into a regular undergraduate degree in STEM disciplines as well as in other disciplines such as education sciences, economics/business sciences and social sciences (Makrakis & Kostoulas-Makrakis, 2015). It consists of core courses, elective courses, and the required capstone course in the three concentration areas, namely, (1) Climate Change, Environment and Society; (2) Climate Change, Economics and Public Policy; and (3) Climate Change, Science and Technology. Each of the core and elective courses will be of six ECTS and the capstone course of 10 ECTS. The capstone course is based on an internship that provides a strong mechanism for integrating academic coursework with practical experience. The amount of the minimum courses to be taken by undergraduate students to qualify for the CLIMASP minor to be awarded in addition to their undergraduate major degree ranges from 45 to 60 ECTS.

Besides formal credential through transcript documentation, students receive the Euro- Arab pass diploma that is adapted from the Europass to certify that they have developed leadership in the field of climate change and sustainability policy. During the CLIMASP project period 240–300 interdisciplinary course syllabi and course modules of 6 ECTS across the 10 partner universities were developed, validated and accredited. In order to institutionalise and support the CLIMASP programme, a Centre for Integrative and Interdisciplinary Studies with an ICT Laboratory has been developed. During the three-year period of the CLIMASP project 100 teaching staff and 3000 undergraduate students were involved in diagnostic, formative and summative assessments.

The CCSAFS (Climate Change, Sustainable Agriculture and Food Security) funded by Erasmus + for the period 2016-2019 and built on RUCAS and CLIMASP projects aims to respond to the needs identified that justify the development of an inter/multidisciplinary MSc programme in CCSAFS, that responds to the Sustainable Develoment Goals, especially SDG 2. The specific objectives are to carry out capacity building for the involved staff from the partner institutions: a) to design post-graduate curricula in line with the Bologna process; b) to design and develop inter/multidisciplinary course curricula at the post-graduate level; and c) to apply innovative and flexible teaching and learning methods. The project has already structured the CCSAFS curriculum, developed a platform for blended learning supplemented by laboratories to support the MSc programme and Centers of Excellence in CCSAFS in each partner country university. The MSc programme in CCCSAFS is worth of 120 ECTS (90 ECTS course work & 30 ECTS thesis) where Suez Canal University from Egypt and Jerash University from Jordan will serve as the main hubs.

As the CCSAFS MSc programme is standardised, students can attend courses in other partner country and EU universities, with credit transfer and the future possibility for earning a dual degree. A blended learning perspective applying a MOOCs approach has been integrated into the CCSAFS course modules. The MOOCs courses will be open, this will highly contribute to lifelong learning, especially to those who will not be registered for the programme.

The CCSAFS curriculum development applied new methods such as participatory or negotiated interdisciplinary curriculum approaches and applying problem-based learning approaches to teaching and learning. This will have a significant impact on the modernisation of the partner countries' institutions through quality teaching. It also promotes the internationalization of the Egyptian and Jordanian partner universities as the MSc programme is compatible with EU standards and the ECTS system as well as the possibility for regional and international student mobility. Through a multi- stakeholder approach, the CCSAFS partnership brings together institutions, NGOs, various stakeholders, experts and students having different social and cultural backgrounds. It will also have a positive impact on building intercultural skills and communication competences as well as strengthen the internationalisation of all partner institutions.

CCSAFS establishes links with various national, regional, European and international organisations, networks and committees, such as the Climate Food and Farming Network (CLIFF), an international research network that helps to build the capacity of young researchers working on climate change mitigation in smallholder farming, the Food and Agriculture Organisation (FAO) of the United Nations; the High Level Panel of Experts on Food Security and Nutrition (HLPE)

Steering Committee and the Committee on World Food Security as well as with the European Commission Joint Programming Initiative on Agriculture, Food Security and Climate Change (FACCE- JPI). Through this kind of networking, we will not only disseminate the CCSAFS world-wide, but also help to strengthen the partnership among the CCSAFS consortium and its external stakeholders.

3. Conclusion: Lesson Learned

Started from the assumption that STEM education has failed to integrate education for sustainability (EfS), the examples of the RUCAS, CLIMASP and CCSAFS projects show that the first lesson-learned is to ensure that the STEM curriculum allows students to be direct participants in creating change in the community. Student learning of STEM becomes meaningful when students are engaged in real-life issues that directly affect everyday life. If properly explored, sustainability issues such as climate change can be a powerful incentive for students to become more engaged in learning because they see the relevance of their learning to their everyday lives.

A second lesson-learned is that STEM students become more engaged in the learning process when they merge theory with practice. Students' placement in community-based practicum with hands-on projects based on their personal interests proved to be very rewarding.

A third lesson-learned is that when the STEM curricula are explored from multiple perspectives, students find meaning in what they learn and become more motivated. The CCSAFS curriculum, for example, gives particular attention to local cultural knowledge. CCSAFS students are introduced to science/green technology/engineering and agricultural science through experiencing the indigenous agricultural knowledge and practices. Merging modernity and tradition may show insightful ways that humanity has learned to survive, and may offer good lessons for building more sustainable ways of living. While interdisciplinary teaching was very seldom a declared practice before the RUCAS project, through its capacity building program, the good practices show a shift away from mono-disciplinary teaching to interdisciplinary teaching and learning.

Last, but not least, a fourth lesson learned from the RUCAS, CLIMASP and CCSAFS projects is that community-based STEM learning encourages students to see their educational development in relationship to what society needs, as opposed to the stereotypical imagery of what STEM is supposed to produce. This has been highly achieved through the participatory curriculum development methodology applied by involving internal (academic oriented) and external (society-oriented) stakeholders from the planning process, to development, implementation and assessment. Curriculum was perceived as process, context and praxis (Grundy, 2003). As a process, reorienting university curricula to address sustainability is not an end, but rather the interaction of teaching staff, students and knowledge. As context, curriculum was contextualized in ways to reflect the local/regional environment. The notion of reorienting a university curriculum as praxis holds that sustainability practice should not focus exclusively on individuals alone nor teaching staff and/or students or the group alone, but also on the way in which individuals and the group create understandings and practices, as well as meaning through critical reflection. EfS was perceived as a means to balance environmental, social, economic and cultural perspectives contextualized through linking the local, the national and global contexts. Engaging students, instructors, administrators, community and public sector, using a variety of pedagogical methods promotes active and participatory learning, sustainability justice as well as active local/global citizenship.

Acknowledgments

This work has been developed within the framework of the CCSAFS (Development of an interdisciplinary MSc programme in climate change, sustainable agriculture and food security) project that has been funded from the European Commission (No. 573881-EPP-1-2016-1-EL-EPPKA2-CBHE-JP (2016-3770-001-001). The content of the paper reflects the views of the authors, and the Commission cannot be held responsible for any use which may be made of the information contained therein.

REFERENCES

American Management Association (2010). *Critical skills survey.* http://www.amanet.org/organizations/2010-survey-critical-skills.aspx

Asikainen, H. (2014). *Successful learning and studying in biosciences: exploring how students conceptions of learning* [Doctoral dissertation, University of Helsinki]. HELDA. https://helda.helsinki.fi/handle/10138/42395

AT21CS (2012). *What are 21st-century skills?* http://atc21s.org/index.php/about/what-are-21st-century-skills/

Clark, B. & Button, C. (2011). Sustainability transdisciplinary education model: interface of arts, science, and community (STEM). *International Journal of Sustainability in Higher Education, 12*(1), 41–54. https://doi.org/10.1108/14676371111098294

Dolmans, D., Loyens, S., Marcq, H. & Gijbels, D. (2016). Deep and surface learning in problem-based learning: a review of the literature. *Advances in Health Sciences Education, 21*(5), 1087–1112. https://doi.org/10.1007/s10459-015-9645-6

Grundy, S. (2003). Αναλυτικό Πρόγραμμα: Προϊόν ή πράξις (Ε. Γεωργιάδη, Trans.). Σαββάλας. (Original work published 1987)

Howie, P. & Bagnall, R. (2012) A critique of the deep and surface approaches to learning model. *Teaching in Higher Education, 18*(4), 389–400. doi: https://doi.org/10.1080/13562517.2012.733689

Makrakis V. (2017). Unlocking the potentiality and actuality of ICTs in developing sustainable–justice curricula and society. *Knowledge Cultures, 5*(2), 103–122. https://doi.org/10.22381/KC5220177

Makrakis, V., & Kostoulas-Makrakis, N. (2013a). Sustainability in higher education: A comparative study between European Union and Middle Eastern universities. *International Journal of Sustainable Human Development, 1*(1), 31–38.

Makrakis, V., & Kostoulas-Makrakis, N. (2013b). A methodology for reorienting university curricula to address sustainability: The RUCAS-Tempus project initiative. In S. Caeiro, W. Leal Filho, C. Jabbour, & U. Azeiteiro (Eds.), *Sustainability assessment tools in higher education institutions* (pp. 23–44). Springer International Publishing Switzerland. https://doi.org/10.1007/978-3-319-02375-5

Makrakis, V., & Kostoulas-Makrakis, N. (2015). A strategic framework for developing interdisciplinary minors on climate change and sustainability policy: The CLIMASP-Tempus example. In W. Leal Filho, W. Azeiteiro, S. Caeiro, & F. Alves (Eds.), Integrating sustainability thinking in science and engineering curricula (pp.103–114). World Sustainability Series. Springer International Publishing Switzerland. https://doi.org/10.1007/978-3-319-09474-8

Makrakis, V., & Kostoulas-Makrakis, N. (2016). Bridging the qualitative– quantitative divide: Experiences from conducting a mixed methods evaluation in the RUCAS programme. *Evaluation and Program Planning, 54*, 144–51. https://doi.org/10.1016/j.evalprogplan.2015.07.008

Marton F., & Säljö R. (1976). On qualitative differences in learning: I Outcome and process. *British Journal of Educational Psychology, 46*(1), 4–11. https://doi.org/10.1111/j.2044-8279.1976.tb02980.x

Marton, F., & Säljö, R. (1984). Approaches to learning. In F. Marton, D. Hounsell, & N. Entwistle (Eds.), *The Experience of Learning* (pp. 39–50). Scottish Academic Press.

Murray, P. E., & Murray, S. A. (2007). Promoting sustainability values within career- oriented degree programs; A case study analysis. *International Journal of Sustainability in Higher Education, 8*(3), 285–300. https://doi.org/10.1108/14676370710817156

Partnership for 21st Century Learning (2011). *P21. Partnership for 21st Century Learning.* http://www.p21.org/index.php

Smith, C. & Watson, J. (2016). *STEM Education and Education for Sustainability (EfS): Finding common ground for a flourishing future* [Paper presentation]. Australian Association for Research in Education (AARE) Conference, Melbourne, VIC, Australia. https://www.aare.edu.au/data/2016_Conference/Full_papers/547_Caroline_Smith.pdf

Sterling, S. (2004). Higher education, sustainability, and the role of systemic thinking. In P. B. Corcoran & A. E. J. Wals (Eds.), *Higher education and the challenge of sustainability: problematics, promise, and practice* (pp. 49–70). Dordrecht, The Netherlands: Kluwer Academic Publishers.

UNESCO (2010). *Education for sustainable development lens: a policy and practice review tool.* UNESCO.

UNESCO (2017). *Education for Sustainable Development Goals Learning Objectives.* UNESCO.

Wals, A. (2012). *Shaping the Education of Tomorrow: 2012 Full-length Report on the UN Decade of Education for Sustainable Development.* UNESCO.

Warburton, K. (2003). Deep learning and education for sustainability. *International Journal of Sustainability in Higher Education, 4*(1), 44–56. https://doi.org/10.1108/14676370310455332

Chapter 17

Mechanical Linkages as Links Between Mathematics and Engineering

Eugenia Koleza[1]

[1]*University of Patras, Department of Primary Education, Greece; ekoleza@upatras.gr*

Abstract: *Our research aims at establishing communication bridges between two apparently disconnected academic communities, the mathematicians' and the engineers', through a common tool: simple and mathematical machines. Mechanical devises may be used as a context for both reinforcing students' spatial abilities and strengthen engineers' construction skills. The main goal is to show the importance of introducing a new register of a concept in mathematics course - the mechanical/ enactif register-in order to improve the students' understanding and learning. This paper reports part of a study investigating the use of a specific mathematical machine, the pantograph for establishing an interdisciplinary culture of apprehending science and mathematical concepts and engineering design.*

Keywords: *STEM, Mathematics, Engineering, pantograph, homothety*

1. The M Component in STEM Education

Last years, bringing STEM Education back to the fore, and given the rapid development of technology, the question about the "value for knowledge" of such an educational approach, become more focus: *"One can almost hear the cry in the halls of state Departments of Education, school district offices, principals' offices, and school corridors: "We do STEM!" But what exactly does that mean? [...] We should ask our leaders exactly what they mean when they use the word "STEM." We deserve more than a generalist blanket response that represents a grouping for funding without specific content or pedagogical substance"* (Shaughnessy, 2013, p.324). In the literature we find many reports suggesting that STEM education provides the context for enhancing the development of mathematical skills. This assumption raises different kinds of questions: What kind of skills is developed? We have development only of skills/abilities or of knowledge also? What kind of abilities? According Goold and Devitt (2014, p. 255), "*The modes of thinking resulting from mathematics education which influence engineers' work performance are: problem solving strategies (identified by 26.4%), logical thinking (26.2%); critical analysis (7.2%); modeling (7.2%); decision making (6.3%); accuracy/confirmation of solution (4.8%); precision/use of rigor (4.6%); organizational skills (4.6%); reasoning (3.6%); communication/teamwork/making arguments (3.2%); confidence/motivation (3.1%), and only (2.2%) is devoted to numeracy and (0.7%) to the use of mathematical tools*". Excepting numeracy, all the others modes of thinking characterize in general the scientific way of thinking. So, the "M" component is for Mathematics or only for Mathematical Literacy/Quantitative Reasoning? Do students acquire a deep understanding on mathematics concepts in the context of STEM education? The National Council of Supervisors of Mathematics (NCSM) has divided basic mathematical **skills** into ten areas: problem solving; **applying** Mathematics to everyday situations; alertness to the **reasonableness** of results; **estimation** and approximation; appropriate **computational** skills; geometry; **measurement;** reading, **interpreting, and constructing tables**, charts, and graphs, using Mathematics to **predict;** and computer literacy. Focusing more on problem solving, Borovik and Gardiner (2006) refer on the **abilities:** to make and use **generalizations**, to offer and use **multiple representations** of the same mathematical object, to approach a problem in different ways, looking for an alternative solution, to utilize analogies and make **connections**. Skills, knowledge, abilities! What is behind the "M" of STEM Education? For the most part of STEM projects, the Mathematics is incidental to the purpose of STEM activities. Just because the numeracy part of mathematics is present in an activity, or students have the occasion of applying general scientific modes of thinking it doesn't mean that students will deepen in math concepts. The STEM activities must be so tightly interwoven with an important math idea/concept, that the students couldn't help but learn about this concept in order to solve the design problem. Deep learning approaches must be correlated with learning experiences that stress "conceptual connections between the content of the learning domain" (Wierstra et al., 2003, p. 514). Nevertheless, the mathematical provision with which many students come to face a real problem in the context of a STEM project, is very often inadequate. Even worse, "*the mathematical thinking processes students develop in their pre-college education may serve as cognitive obstacles students must overcome in order to develop the design thinking skills which are critical for engineering practice broadly as well as for creativity and innovation*" (Tolbert & Cardella, 2013, p. 23). Mathematics is often mentioned as underpinning the other disciplines of STEM because it serves as a language for Science, Engineering, and Technology. What is the language of Maths? Just the symbols and the expressions, or the deep knowledge of concepts and the math way of thinking also? Furthermore, stating that Mathematics underpins the other disciplines sets Mathematics up in a supporting role in integrative STEM education contexts, and less as an enabler or imperative for the advancement of understanding of concepts in other disciplines. The pressure for competition and performativity, have allowed Engineering and Technology education to emerge as a prioritized necessity: a necessity to 'put things to work' quickly, creatively, and efficiently. It seems that Mathematics are not highly valued in performative knowledge! NCTM president J. M. Shaughnessy complains for the fact that subordinate role of Maths in STEM education: *"Mathematics is paramount, Mathematics is primal, Mathematics is the most important STEM discipline. The other three disciplines are fundamentally dependent on the strong mathematical preparation of our students"* (Shaughnessy, 2013, p. 324). So, how can we emphasize in a STEM context the understanding of Mathematics, while ensuring applicability? An idea: highlighting the relationship between Mathematics and Engineering.

2. Connecting E&M in STEM Education

The tools of Mathematics is indispensable to Engineering. Mathematical thinking, and more specifically mathematical modeling- as one of the key mathematical thinking skills-is needed for the analysis and design of engineering structures and systems. Nevertheless, in mathematics education, as reported by Tolbert and Cardella (2013 students are often taught to follow a linear, methodical process to reach the one best solution. As a result, when students are given tasks to complete they seem to have a limited understanding of how to develop appropriate mathematical models to help them solve the problems. Many researchers and academics in engineering departments testify/ (report about) to the declining mathematical skills of engineering students and the overwhelming preference for minimal-mathematics. The **synergy** of Mathematics and Engineering in the context of a STEM pre-college Education could contribute at improving the mathematics preparedness of new engineering undergraduates. The idea is not really new: using Engineering for mathematical purposes dates back to the ancient Greece, 2300 years before. During a two-day world conference held in Spring 2013 at the Courant Institute of Mathematical Sciences, New York University, entitled "Archimedes in the 21st Century", the conference talks were divided into three categories: Archimedes the Mathematician/Geometer, Archimedes the Scientist, and Archimedes the Engineer, the fields of Mathematics, Science, and Engineering each claiming Archimedes as one of their own. Archimedes, is nowadays better known as an engineer because of the "engines" with which he thwarted the Roman siege of Syracuse (214-212 B.C). Certainly, what was considered a machine two thousand years ago differs considerably from our present ideas. Machines were invented mainly to help pupil for the moving of heavy weights. Hero of Alexandria summarized the practice of his day by naming the five simple machines for moving a given weight by a given force. In the eighteenth century, scientists begin to be interested mainly to the idea of modifying motion than constructing a machine for practical reasons. During this period appear two great figures: Euler and Watt. Euler's work was in applying analysis to the science of movement: The concept that planar motion may in general be described by a translation of a point and a rotation about this point, appears in 'Mechanica sive motus scientia analytice exposita' (1736-1742). Thus, kinematic analysis stems from Euler, while the synthesis of movement from Watt. *"The investigation of the motion of a rigid body may be conveniently separated into two parts, the one geometrical, the other mechanical"* (Euler, 1775). In Kinematics (of Mechanisms) Analysis, given a mechanism, the task is to analyze its motion-displacement, velocity, and acceleration. In Kinematics synthesis, given a desired motion, the task is to develop a mechanism that meets the requirements. Except Euler, his contemporaries D'Alembert and Kant treated also motion in a purely geometric manner ("geometric motion" for Carnot), disregarding the causes of motion. The term cinematique/ kinematics (derived from the greek word κινηματική) was first introduced by André-Marie Ampere in the "Essai sur la philosophie des sciences" (1834): *"This science ought to include all that can be said with respect to motion in its different kinds, independently of the forces by which it is produced. It should treat in the first place of spaces passed over, and of times employed in different motions, and of the determination of velocities according to the different relations which may exist between those spaces and times"*. The science of kinematics, the science of pure motion, is an area of study where motion is analyzed without regard to the forces that cause it. It is essentially geometry in motion. Understanding the changing geometry of a linkage needs kinematic analysis. Kinematic synthesis or linkage synthesis entails conceiving an arrangement of assembled bodies and determining their sizes to get the desired motion. The simpler propositions of kinematics are based on geometrical principles, since they deal with the ideas of position and space. The introduction of the ideas of time and consequently of velocity and acceleration, extends the scope of kinematics beyond the limits of pure geometry. Two classes of problems arise: the first dealing with the position and motion of a particle, and the second treating of similar questions relating to rigid bodies. The motion of non-rigid bodies is of course of a far more complex nature, because it cannot be investigated apart from the forces acting on them, and its consideration falls within the province of Kinetics, rather than within that of Kinematics. The revolutionary idea of creating a classification of the mechanisms depending on their purpose, is attributed to Willis (1841), while modern kinematics-using symbols for the description of kinematic chains instead of pictures- has its beginning in 1875 with Reuleaux, who was able to show that mechanisms with different physical appearance could be kinematically identical. Mathematicians, and especially geometers discovered that any algebraic curve-of any order- can be described by a link motion. From 1877 when first appeared the lecture on Linkages (: How to Draw a straight line) by Kemp (Kempe has demonstrated that there exists a connecting rod mechanism to trace any algebraic curve), till the Bryan and Sangwin edition "How Round is your Circle? Where Engineering and Mathematics Meet" (Bryan & Sangwin, 2011), teaching Mathematics through a

Machine-based approach was **a nascent idea**, which has never really applied in Mathematics Education, apart from some exceptions. It is the case of Italian researchers (Bartolini & Martignone 2015) who have recommended using artifacts and contexts of geometric practice which employ mechanical or jointed models of drawing and tracing machines as a school's way to generate complex mathematical ideas or notions. So, while the synergy of Engineering and Mathematics is not a new idea, the way we can introduce it in the education, is. Among the few contributions to the field we mention the research by R.Gras (1983), Métrégiste (1984), and Brousseau (1998) who used in "Problems on the Didactic of Decimals", the classic pantograph that performs homothéties. More recently, we see appearing the research by Isoda et al. (1998), Vincent (2003), and English et al. (2013) An extremely valuable proposal for creating a connection between Engineering and Mathematics (different that the machine-proposal) has been put forward by researchers of School of Education, Purdue University, through the development of real-world engineering tasks, called model eliciting activities (MEAs; Moore & Diefes-Dux, 2004).

3. Our Research: Engineering Mathematics

In our project, the idea we elaborate is the one of reverse engineering: reverse engineering is the process of discovering the technological principles of a device, object, or system through analysis of its structure, function, and operation. It often involves taking something (e.g., a mechanical device or electronic component) apart and analyzing its workings in detail in order to see how a product works, what it does, what components it consists of, to be used in maintenance. By carefully disassembling, observing, testing, analyzing and reporting, engineers can understand how something works and suggest ways it might be improved.

Our research is not just about the structure and function of a mechanical linkage, but about the mathematical concepts embodied into the linkage. In this project, mathematics inquiry, meets the Engineering Design, and «stands against the misconception of replacing mathematics inquiry with engineering design, a mistake made by many school teachers in their early stage of implementing STEM (Tinga 2015, p.60). Here we report some findings from a preliminary research with a pantograph used by grade 9 students. Elsewhere we had presented our research with pantograph with grade 10 students (Siopi & Koleza, 2017). In the "pantograph project" our aim is to investigate the potential of a mechanical linkage-the pantograph-in creating an environment for year 9 students to invent (so as to be able –later- to recognize) the concept of homothety, while arguing and constructing geometrical justifications. More precisely we aimed at investigating whether: 1) Students are able to: recognize mathematics in a linkage and use mathematics for constructing linkages in a context of reverse engineering 2) The construction of a linkage under specific constraints (given ratio between the rods) follow the same steps as an engineering design.

The pantograph we worked with is a special kind of a chain of four links connected by turning pairs whose axes are parallel. When the links of this chain are of unequal lengths the smallest is called the crank, and since the four links form a quadrilateral. Reuleaux has called this chain the quadric crank-chain. In quadric crank-chains it will be convenient to distinguish between links having a partial turning movement and those, which can execute complete rotations relatively to the fixed link in the chain. The former links will be called levers, the latter cranks.

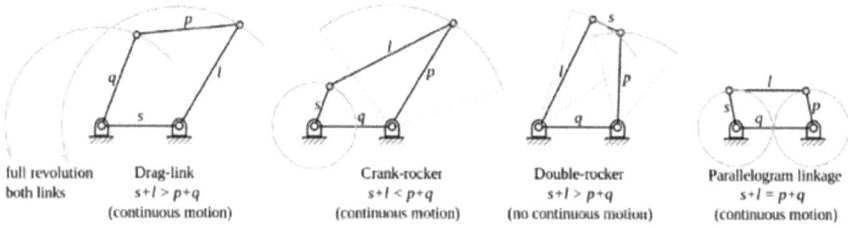

full revolution both links	Drag-link	Crank-rocker	Double-rocker	Parallelogram linkage
$s+l > p+q$ (continuous motion)	$s+l < p+q$ (continuous motion)	$s+l > p+q$ (no continuous motion)	$s+l = p+q$ (continuous motion)

Figure 1: Grashof condition

A quadric crank-chain in which opposite links are of equal lengths is employed (with the addition of a fifth fixed link) for copying purposes under the name of a pantograph. Invented by the German astronomer Christolph Scheiner (1603) appears in different forms. The most commercial one- is pictured in figure 2. The pantograph in geometry teaching can be used (as in our case) for the intuitive discovery of the homothety transformation.

Four equal rods are hinged by pivots at A, B, C, P, with **OA = AP and PC = P'C = AB.** The instrument is fastened to a plane surface, by a **pointed pivot at O**. Then if pencils are inserted at P and P', and P is made to trace a figure F, P' will trace the figure F' obtained from F by the homothety. The justification of the machine's function results by showing that (1) APCB is a parallelogram, (2) O, P, P' are collinear, and (3) OP'/OP=OB/OA=constant. (Eves 1995, p.108)

PC=AB and **BC=BP'-CP'=BO-AB=AO=AP** so APCB is a parallelogram **(1): AP//BP'**

Angles P1+P2+P3=O+B+P'=180 so **O, P, P' are collinear (2)**

Triangles OAP & OBP' are similar so OP'/OP=OB/OA=k (constant) (3).

For every point of figure F, we can find a corresponding point on F' so as OF'/OF=k. So, the movement of the pantograph creates homothetic figures. An intuitive explication of the pantograph's function can be given observing figure 3.

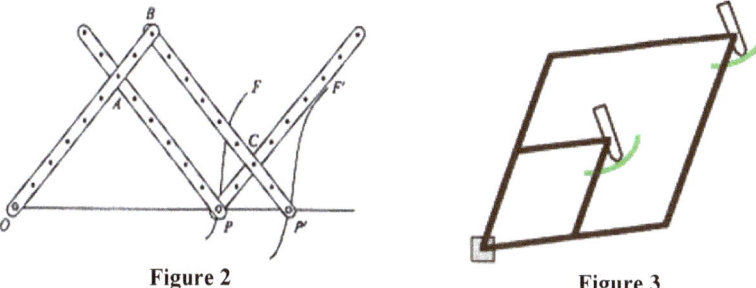

Figure 2   Figure 3

A precise justification can be given also by using vectors.

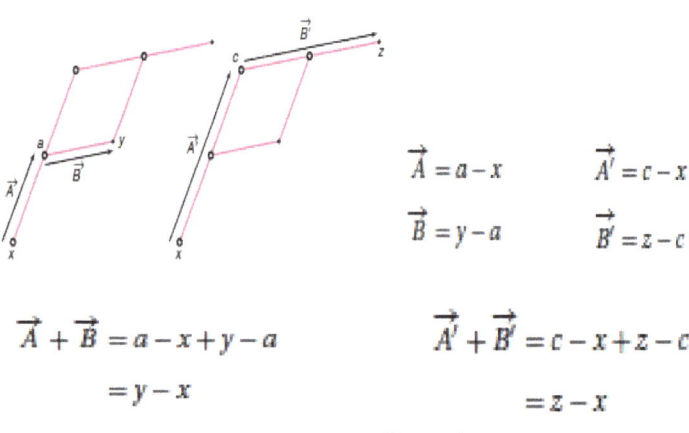

$\vec{A} = a - x$ $\quad\quad$ $\vec{A'} = c - x$

$\vec{B} = y - a$ $\quad\quad$ $\vec{B'} = z - c$

$\vec{A} + \vec{B} = a - x + y - a$ $\quad\quad$ $\vec{A'} + \vec{B'} = c - x + z - c$

$= y - x$ $\quad\quad\quad\quad\quad\quad\quad\quad$ $= z - x$

Figure 4

If we place the origin of a coordinate system at x = (0,0), then z − x = z and y − x = y, the above equation says that the coordinates of z are always double the coordinates of y in this coordinate system. So every drawing traced by y is traced at twice the size by z.

One of the many applications of the pantograph is his function as a "double lever mechanism" (Watt's indicator mechanism). It consists of four links, is shown in Fig. 5. The four links are: fixed link at A, link AC, link CE and link BFD. It may be noted that BF and FD form one link because these two parts have no relative motion between them. **The links CE and BFD act as levers**. The displacement of the link BFD is directly proportional to the displacement of the tracing point E at the end of the link CE traces out approximately a straight line.

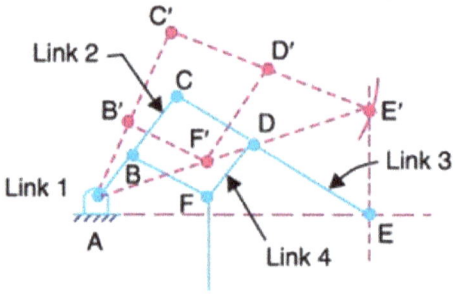

Figure 5

4. Methodology

Participants were two groups of three high school students (15 years old). Two girls and four boys, who had no previous formal experience to deductive reasoning. We completed our project in 5 meetings of 2 hour each, following the participant observation method. After the fifth meeting, we had a personal, focused interview with each student, where we asked him or her to explain their answers. We recorder and videotaped all meetings. Briefly, the structure of the project was as follows: **Phase 1** (2 meetings/1th and 2th): (Perception of the machine) *How is the machine made? What does the machine do?* **Phase 2** (1 meeting/3nd): (Justification of perceptions) *How is the machine made?* **Phase 3** (1 meeting/4th): (Construction under restrictions) *How the machine could be in order to?* **Phase 4** (1 meeting-two weeks later/5th) *Why does the machine do that?*

4.1. Comments on Students' Work

Aiming at highlighting the connection between Engineering and Mathematics, and because of the restriction of space, we will comment only on the final meeting.

5th **Meeting:** *Why does the machine do that?*
Activity 4: *Graph A makes a segment (e). Complete the figure appropriately with the drawing written by B.* (Figure 6)

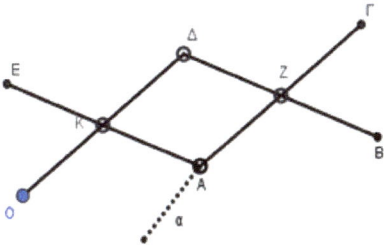

Figure 6

Pantelis: We take the triangle OKA, and <u>we lower one corner</u>, A, <u>pull</u> the line 1 cm. Triangle OΛB is 4 times bigger (though he shows the perimeter. He means that the perimeter is double, but confused by the image he refers to the area). So the parallel line that lowers B will be double. (Figure 7)

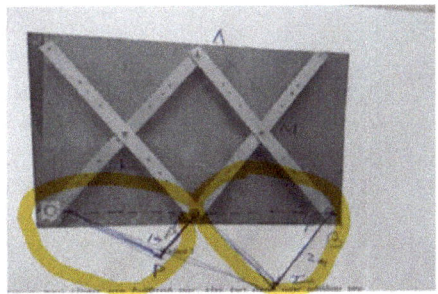

Figure 7: The work of Pantelis in activity 4

Researcher: Why not 4 times bigger?
Pantelis: ... O, A, B are in a straight line and A is in the middle (: he abandons the triangles and focus only to the relation OB=2OA).
Researcher: How it is related to the fact that the machine produces a double length segment?
Pantelis: ...
Researcher: You told me that AP and BΣ are parallel lines. Can you locate these parallels as part of a more complex shape? This relationship you've identified OB=2OA does it exist elsewhere in your pantograph?
Pantelis: ΛB=2ΛM [*by the construction of the tool*]. **I don't know...I don't know.**

Activity 5: *Design a pantograph that can triple a segment. Justify why he can do that.*

Pantelis: E <u>has a weight for one piece, while Z has a weight for three and drops three</u>. E is at the corner of the small triangle and Z at the corner of the big one and drops three (Figure 8).

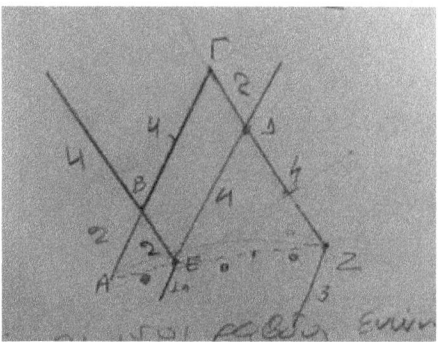

Figure 8

5. Final Comments

Research with pantograph has been documented by Bartolini and Martignone (2015). In their case, participants were teachers from secondary schools. In their conclusions they report that: *"In order to solve this problem teachers had to clearly understand **how the transformation properties are embodied in the artefact structure** and to be able to modify the artefact maintaining the main features of the transformation: the fixed and tracers points remain aligned and at a proportional distance during the movement.* In our case we worked with 8th grade secondary students, who had a very limited learning experience on similitude, and absolute none experience on the concept of homothety as it has been excluded from the curriculum the last 8 years. The teaching of the similitude mainly concerns, understanding of the concept of similar shapes and the constructions that the children are asked to deal with, are usually placed in a horizontal position, i.e., the sides are parallel to the sides of the object on which the construction is done. As a result most students develop a stereotype view of the geometrical shapes, which is very affected by the intuitive rules. The students in our project managed to recognize the similar triangles in the structure of the machine, in a rather static way. They were unable to reconstruct the pantograph under the new conditions: ratio 1:3. The synthesis process proved very difficult due not only to a lack of understanding of the transformation properties embodied in the machine, and the prevalence of the prototype image concerning the similitude of triangles (similar triangles in a horizontal position), but, also, to their inability to make the drawing of the machine on the paper, fact that deserves further investigation, and it was mentioned already by Martignone and Antonini

(2010): "[…] what lets Anna to do the discovery of the transformation incorporated in the machine, is the drawings analysis more than the machine structure".

In the final activity, where they were asked to give argumentations that justify why the pantograph does a homothety, they saw the points movement and the whole pantograph (in a wrong way) as a beam bending under the press of weights «concentrated» in the points E and Z, than as a transformation tool.

The fact that the students look for the interpretation of the machine's function in nature rather than in mathematics the interpretation rather than mathematics, although it opens up perspectives for the correlation of these two approaches, but also it raises concerns about the level of understanding of mathematics. The basic questionσ that emerged through the analysis of the observations were: What are the appropriate questions that will lead students who have been taught a concept to discover it through the operation of a machine? Is the pantograph the best mechanical approach of the homothety concept? According to Bartolini and Martignone (2015), *"in order to proof why the artefacts does that, they have to use arguments linked to the articulated system […] the argumentations based on movement lead to further argumentations that explain the motion through the structure of the articulated system, and a cognitive unity may or may not occur"*. Does this mean that the pantograph experiment can be apprehended only with elder students? Does the explanation of the motion through the structure of the machine, could be understood if we highlight best the double nature of the pantograph: as a math machine and a physical system? Our research is geared to answering these questions.

REFERENCES

Bartolini, M. G., & Martignone, F. (2015). The Use of Concrete Artefacts in Geometry Teacher Education for Secondary School. In J. Huang, Y. Li, & W.-H. Poon. Proceedings of the *Hong Kong Mathematics Education Conference 2015* (pp. 24–34). Hong Kong Association for Mathematics Education.

Borovik, A., & Gardiner, T. (2006, 22–28 July). *Mathematical abilities and mathematical skills* [Paper presentation]. World Federation of National Mathematics Competitions Conference 2006, Cambridge, England. http://eprints.maths.manchester.ac.uk/839/1/abilities.pdf

Brousseau, G. (1981). Problemes de didactique des decimaux [Problems in teaching decimals]. *Recherches en Didactiques des Mathematiques, 2(*1), 37–125

Bryant, J., & Sangwin, C. (2011). *How round is your circle? Where engineering and mathematics meet.* Princeton University Press.

English, L. D., Hudson, P. B., & Dawes, L. A. (2013). Engineering based problem solving in the middle school: Design and construction with simple machines. *Journal of Pre-College Engineering Education Research, 3*(2), 1–13.

Eves, H. W. (1995). *College geometry.* Jones & Bartlett Learning.

Goold, E., & Devitt, F. (2014). Mathematics in engineering practice: tacit trumps tangible. In B. Williams, J. Figueiredo, & J. Trevelyan (Eds.), *Engineering Practice in a Global Context: Understanding the Technical and the Social* (pp. 245–264). CRC Press.

Gras R. (1983), Instrumentation de notions mathématiques, un exemple: la symétrie. *Petit X, 1,* 7–39.

Isoda, M., Matsuzaki, A., & Nakajima, M. (1998). Mathematics inquiry enhanced by harmonized approach via technology: A crank mechanism represented by LEGO and graphing tools. In H. S. Park, Y. H. Choe, H. Shin, & S. H. Kim (Eds.), *Proceedings of ICMI-EARCOME 1:*

<blockquote>
The First International Commission on Mathematical Instruction, East Asia Regional Conference on Mathematics Education (Vol. 3, pp. 267–278). Korea Society of Mathematical Education.
</blockquote>

Martignone, F., & Antonini, S. (2010). Students' utilization schemes of pantographs for geometrical transformations: a first classification. In V. Durand-Guerrier, S. Soury-Lavergne, & F. Arzarello (Eds.), *Proceedings of the Sixth Congress of the European Society for Research in Mathematics Education* (pp. 1250–1259). INSTITUT NATIONAL DE RECHERCHE PÉDAGOGIQUE.

Métrégiste, R. (1984). La géométrie des transformations: une approche en classe de 4e et 3e. *Petit X, 4*, 35–71.

Moore, T., & Diefes-Dux, H. (2004). Developing model-eliciting activities for undergraduate students based on advanced engineering content. In IEEE (Ed.), *Proceedings 2014 IEEE Frontiers in Education Conference* (pp. F1A-9). https://doi.org/10.1109/FIE.2004.1408557

Shaughnessy, M. (2013). Mathematics in a STEM context. *Mathematics Teaching in the Middle School, 18*(6), 324. https://doi.org/10.5951/mathteacmiddscho.18.6.0324

Siopi, K. & Koleza, E. (2017). Tool-based argumentation. In Dooley, T., & Gueudet, G. (Eds.). *Proceedings of the Tenth Congress of the European Society for Research in Mathematics Education* (pp. 259–266). DCU Institute of Education and ERME.

Tinga, Y.-L (2016). STEM from the perspectives of engineering design and suggested tools and learning design. *Journal of Research in STEM Education, 2*(1), 59–71.

Tolbert, M. D. A., & Cardella, M. E. (2013, June 23–26). Early work for the Mathematics as a Gatekeeper to Engineering Project: A Review of Informal Learning, Engineering and Design Thinking Literature. 120th American Society for Engineering Education (ASEE): Annual Conference & Exposition, Atlanta, GA, USA. https://peer.asee.org/early-work-for-the-mathematics-as-a-gatekeeper-to-engineering-project-a-review-of-informal-learning-engineering-and-design-thinking-literature.pdf

Vincent, J. (2003). *Mechanical linkages, dynamic geometry software, and argumentation: Supporting a classroom culture of mathematical proof.* [Unpublished doctoral dissertation]. University of Melbourne.

Wierstra, R. F. A., Kanselaarl, G., Van der Linden, J. L., Lodewijks, H. G. L. C., & Vermunt, J. D. (2003). The impact of the university context on European students' learning approaches and learning environment preferences. *Higher Education, 45*(4), 503–523.

www.ingramcontent.com/pod-product-compliance
Lightning Source LLC
Chambersburg PA
CBHW062026290426
44108CB00025B/2798